Bubbles, Panics, and Crowds

A History of Financial Manias

Joseph Trek

Published by NobleTrex Press

For permissions and other inquiries, write to:

P.O. Box 3132, Framingham, MA 01701, USA

Contents

Introduction

In the spring of 1720, Sir Isaac Newton-the man who unlocked the secrets of gravity, invented calculus, and fundamentally reshaped our understanding of the universe-decided to sell his shares in the South Sea Company. He pocketed a handsome profit of £7,000, a fortune at the time, and congratulated himself on a prudent financial maneuver.

But then, something irritating happened. The stock kept going up.

He watched as his friends, neighbors, and people distinctly less intelligent than himself continued to get rich. The narrative on the streets of London was intoxicating; the South Sea Company held a monopoly on trade with Spanish America, and the potential for wealth seemed infinite. The social pressure-the sheer agony of watching the herd gallop toward paradise without him-became unbearable. Newton re-entered the market near the peak, investing nearly everything he had.

When the bubble burst a few months later, the smartest man of his era lost £20,000, the equivalent of millions of dollars today. He famously remarked, "I can calculate the motions of heavenly bodies, but not the madness of people."

If the father of modern physics couldn't insulate himself from the seductive pull of a financial mania, what chance do the rest of us have?

That is the question at the heart of this book. We like to imagine that money is a matter of mathematics-a clean world of spreadsheets, discounted cash flows, and rational actors making calculated decisions. We convince ourselves that with enough data, the right software, or the perfect algorithm, we can tame the beast of uncertainty.

But history tells a different story. Financial markets are not merely calculators; they are stadiums of human emotion, battlegrounds of psychology, and, occasionally, theaters of the absurd.

This book is about what happens when the theater catches fire.

You are holding a history of financial manias, but it is not a collection of dusty dates and antiquated anecdotes. It is a study of human nature. Whether the asset in question is a tulip bulb in the Dutch Republic of the 1630s, a railway share in Victorian England, a suburban house in 2007, or a meme stock driven by a Reddit forum in 2021, the pattern is identical. The technology changes, the delivery mechanisms speed up, and the jargon evolves, but the software running in our brains remains stubbornly, dangerously prehistoric. We are wired to seek safety in numbers, to fear missing out more than we fear losing money, and to believe the most compelling story rather than the most accurate data.

Why does this matter to you? Because we are living through an era where the lines between investment, entertainment, and gambling have blurred beyond recognition. We carry the stock market in our pockets. We receive push notifications that trigger the same dopamine receptors as a slot machine. The barriers to entry have collapsed, allowing global crowds to form and disperse in seconds. Understanding the anatomy of a bubble is no longer just the concern of Wall Street historians or central bankers; it is a survival skill for anyone trying to preserve their purchasing power in a volatile world.

To navigate this landscape, we first have to look under the hood of the financial system. We tend to view the market as a monolith, but it is actually a delicate ecosystem of plumbing-liquidity, credit, and settlement. When you strip away the noise, you find that the "engine" of finance runs on a mixture of cold arithmetic and hot air. We will examine how prices are actually formed, not in the idealized curves of an economics textbook, but in the messy reality of bids and asks. We will look at the concept of "value" versus "price," and why the two can diverge so violently that they seem to inhabit different dimensions. This foundational knowledge is crucial because you cannot understand why a car crashes if you don't understand how the brakes work.

Once we understand the mechanics, we must confront the mind. The most dangerous asset in any portfolio is the investor's own ego. There is a specific, recurring psychological state that fuels every mania: the suspension of disbelief. We will explore how narratives-"This time is different," "The internet changes everything," "Housing never goes down"-overpower fundamentals. We will dissect the social feedback loops that turn individual optimism into collective euphoria. This is where the concept of "reflexivity" comes in-the idea that rising prices change the fundamentals, which in turn justify higher prices, creating a self-fulfilling prophecy that feels like magic until the moment it stops.

But bubbles are not just psychological phenomena; they are credit events. Optimism alone cannot inflate a bubble; it requires fuel. That fuel is leverage. We will walk through the "Minsky Cycle," named after the economist Hyman Minsky, who famously argued that stability is destabilizing. It is a profound paradox: long periods of economic calm encourage investors to take on more risk and debt, believing the good times are permanent. This accumulation of debt creates a fragile structure where a minor event-a missed earnings report, a small bank failure, a geopolitical rumor-can trigger a catastrophic collapse. We will trace how this leverage has mutated over the centuries, from simple promissory notes to the opaque, trillion-dollar derivatives and shadow banking systems that nearly broke the world in 2008.

The history of finance is, in many ways, a history of forgetting. Every generation believes it has solved the riddle of risk. In the 1920s, they believed they had reached a "permanently high plateau" of prosperity. In the 1990s, they believed the "Great Moderation" had tamed the business cycle. In the 2000s, they believed financial engineering had distributed risk so perfectly that it had effectively vanished.

Each time, the market humbled them. We will revisit these echoes of the past, not to mock our ancestors, but to recognize ourselves in them. We will see how the Victorian railway mania was the 19th-century equivalent of the dot-com boom-a technological revolution that was fundamentally correct about the future but disastrously wrong about the price. We will look at the "Lost Decades" of Japan to understand how a bubble's shadow can stretch for a generation.

We must also confront the modern era, where the game has changed in speed and scale. The collapse of the housing market in 2008 revealed that the financial system had become so interconnected that a default in a suburb of Las Vegas could bankrupt a municipal council in Norway. We will examine the role of the "Rescuers"-the central banks and governments that step in to stop the bleeding. This intervention creates its own complex dilemma: moral hazard. If the government promises to catch you when you fall, why would you ever stop dancing on the ledge? The post-2008 world has been defined by cheap money and aggressive policy, creating an environment where asset prices are tethered as much to the decisions of the Federal Reserve as they are to corporate profits.

Finally, we will turn our gaze to the present and the future. The rise of cryptocurrencies, the phenomenon of meme stocks, and the gamification of trading suggest that the crowd is becoming more powerful, more aggressive, and more unpredictable. We are seeing manias born and dead within the span of days, driven by algorithms and influencers. The principles of the past still apply, but the tempo has shifted from a waltz to a rave. We need new heuristics to spot the warning signs in a world of viral finance.

This book will not give you a formula to time the market top. Anyone who claims they can do that is either a liar or about to lose their shirt like Isaac Newton. Instead, this book offers a framework. It provides a lens through which to view the chaos. When you understand that a sudden spike in asset prices is often just a transfer of wealth from the impatient to the patient, you become less susceptible to the fear of missing out. When you recognize that leverage is a double-edged sword that cuts deepest when the market feels safest, you become more skeptical of "easy" returns.

We are about to embark on a tour of delusion, greed, and catastrophic error. It is a story of scoundrels and visionaries, of bankruptcies and bailouts. But ultimately, it is a story about the tension between our desire for wealth and the reality of risk.

The market is a mirror. It reflects our collective hopes, our deepest fears, and our infinite capacity for self-deception. If we look closely enough at the bubbles of the past, we might just see our own reflection staring back. Let's begin.

Chapter 1

Foundations of Markets, Value, and Risk

Before we can understand why financial markets occasionally lose their minds, we must understand how they function when they are sane. Markets are not just chaotic casinos; they are sophisticated engines built to solve problems of allocation, time, and trust. In this chapter, we will open the hood to see how prices are actually discovered, why a dollar today is worth more than a dollar tomorrow, and how the invisible plumbing of liquidity keeps the system flowing-until it suddenly doesn't. We finish by examining the most potent drug in finance: credit, the amplifier that turns manageable fluctuations into catastrophic booms and busts.

1.1 The Engine of Finance: Money, Markets, and Prices

A bottle of water costs about the same as a candy bar—until you're stranded on a highway in the desert with a dead radiator and the sun turning the hood of your car into a skillet. Then the water isn't "a dollar." It's relief. It's options. It's time. If someone walked up with a cooler and said "twenty bucks," you might pay without negotiating, and then (later, in the shade) feel faintly ridiculous about it. That embarrassment is a clue. Prices aren't moral judgments and they aren't engraved in nature. A price is a truce struck in a particular moment between what a buyer can't stand to go without and what a seller can't stand to part with.

Markets exist because we're always trying to make that truce less chaotic. Without markets, everything is negotiation: every transaction a little duel of ignorance where each side tries to guess what the other side *would* accept. Markets reduce the guesswork by concentrating the argument in one place. They turn millions of private hunches-need, greed, fear, patience-into a public number. That number looks clean. The process that creates it is not.

Here's the first unsettling fact: the price of something is not primarily about what it *is*. It's about how badly it's wanted *right now* relative to how many units can be offered *right now*. That's why you can see the same house sell for $700,000 in one year and $520,000 a year later without the kitchen getting worse or the roof caving in. The house didn't change; the crowd did. Financing costs changed. Local job prospects changed. A few buyers vanished. A few desperate sellers appeared. "Value" can be slow-moving. "Price" can sprint.

Think of the simplest market imaginable: a farmers' market at closing time. There are baskets of peaches left, and a few shoppers lingering. Early in the morning, peaches were scarce and shoppers were plentiful; the farmer could hold firm. Late in the day, the power flips. The farmer now owns a fragile asset with a deadline: tomorrow those peaches are bruises. Shoppers can smell the deadline. The same peaches, same farmer, same stall—different clock, different balance of bargaining power, different price. Supply and demand sound like a diagram from a high school textbook until you watch a real human being try to avoid throwing food away.

Financial markets add rocket fuel to that basic story: they let you trade claims on future cash flows and future resale possibilities at enormous speed. In a grocery store, you don't usually buy milk because you think it will be worth more next week. In markets, that's not just common—it can become contagious. And when it becomes contagious, you start to see the outline of a bubble long before the headlines are brave enough to name it.

What a price really is: the last agreement, not the eternal truth

When people say "Apple is trading at $X" or "Bitcoin is at $Y," it sounds like a statement about the asset itself, as if the number is an intrinsic property like weight. It isn't. That quote is the last recorded handshake. It is the most recent moment when a buyer and seller agreed to stop arguing.

This matters because the last handshake can be strangely unrepresentative. If a stock has millions of shares, but only a tiny fraction actually changes hands today, the "market price" is still determined by the tug-of-war at the margin—by the people *willing* to trade at that moment. A crowd can be euphoric or panicked even if most owners are sitting quietly, hands in pockets, refusing to sell. Prices are set by the *active* minority. In calm times, that's fine. In manias and panics, it's a trap: the loudest participants become the authors of reality.

A second unsettling fact follows: liquidity—how easy it is to trade without moving the price—quietly shapes everything. A suburban house is "worth" something, but it's not a continuous auction. You can't click "sell" and be done. A share of a large public company is different: it's designed to be traded, and the ease of trading makes the price feel objective. That feeling is seductive. It's also why market prices can become a kind of social proof: a number on a screen that looks like a verdict from the universe rather than the output of a crowd.

If you want a clean mental model, try this: price is the answer to a narrow question. Not "What is this worth in some timeless sense?" but "What is the cheapest price at which someone is willing to sell *now*, and the highest price at which someone is willing to buy *now*?" The moment you add *now*, you add human constraints—deadlines, margin calls, embarrassment, rumors, pride, rent due on Monday. "Now" is where finance becomes psychology.

The auction you never see: bids, asks, and the spread

Most modern financial markets are a high-speed auction running in the background. You don't see people shouting and waving paper tickets anymore, but the logic is the same. Buyers submit *bids* (prices they're willing to pay). Sellers submit *asks* (prices they're willing to accept). When a bid meets an ask, a trade happens.

The distance between the best bid and the best ask is called the *spread*. On a calm day in a heavily traded stock, the spread can be a penny. In a thinly traded small stock, it might be ten or twenty cents. In a crisis, spreads can widen like a road cracking during an earthquake. That widening isn't just an inconvenience; it's a signal that the market is struggling to agree on reality. When people say "the market is liquid," what they often mean is "the spread is tight and I can trade without feeling mugged."

The spread is also a reminder that "the price" isn't a single number. At any moment there are at least two: the price you can sell for and the price you can buy for. The difference is friction. It's the toll you pay for immediacy. If you ever wonder why day trading is harder than it looks, start there: you begin every round-trip trade by stepping into a small hole. You have to climb out before you make a profit.

Spreads do something else: they discourage casual flipping and reward patience. A tight spread makes it cheap to change your mind. A wide spread punishes indecision. During bubbles, spreads often compress as activity explodes and confidence feels effortless—one reason manias can accelerate. During panics, spreads often expand as fear makes everyone demand a larger safety buffer—one reason crashes can feel like the floor disappeared.

Three orders, three personalities: you choose what you fear

Behind every trade is a decision about what kind of uncertainty you can tolerate. Retail trading platforms make it feel like you press a button and the market obeys. In reality, you are choosing between different kinds of risk, whether you know it or not.

A *market order* says, essentially: "Just get me in or out." It prioritizes execution. The cost is that you surrender control over the exact price, especially when the market is moving quickly or when there aren't many orders sitting nearby. In calm conditions, the difference might be tiny. In a sudden vacuum—an earnings surprise, a scary headline, a flash of panic—it can be shocking.

A *limit order* says: "Only trade if I can get *this* price or better." It prioritizes price. The cost is that you might not get filled at all. Limit orders are how patient traders avoid paying the spread more than necessary, but patience has its own risk: the market can run away without you, or it can hit your limit for an unpleasant reason—because new information arrived and everyone now thinks your price was too generous.

A *stop order* is the emotionally revealing one. It says: "If the price gets bad enough, get me out." It's a seatbelt that can turn into a catapult. When the stop triggers, it typically becomes a market order, meaning it can execute at a much worse price if the market gaps down or cascades. In calm times, stops feel like discipline. In violent times, they can become gasoline, adding forced selling precisely when selling is already overwhelming.

These mechanics aren't trivia. They're part of the engine. Markets don't just reflect human emotion; they *translate* it into action through rules. A crowd is one thing. A crowd wired to automated triggers is something else. In bubbles and panics, the plumbing of order types often determines how quickly a story becomes a stampede.

Two ways to own: the orchard and the hot potato

Now for the distinction that separates investing from speculation. The words get tossed around as insults or compliments, but they're best understood as different reasons for buying.

Imagine you buy an orchard. If you're an *investor*, you care about the harvest. You want the fruit—the cash that comes out of the asset over time. You might sell someday, but selling is not the point. The point is that the orchard produces something valuable even if nobody is excited about orchards this year. You can be wrong about the resale price and still do fine if the harvest is strong.

If you're a *speculator*, you might buy the same orchard for a different reason: you believe someone else will pay more for it soon. The orchard's fruit is a footnote. The real product is the ability to resell. Speculation is not automatically foolish—markets need people willing to take views on price—but it is structurally dependent on the crowd's future behavior. Speculation is the art of guessing what other people will guess.

That difference sounds small until you live through a drawdown. When prices fall, the investor asks, "Has the orchard's harvest changed?" The speculator asks, "Where did the buyers go?" Those questions lead to very different levels of panic.

There's a useful rule of thumb: if you cannot explain what cash flow you expect an asset to generate for you (rent, dividends, interest, business profits) and you're relying primarily on resale to make money, you are speculating—whether or not the asset is a stock, a condo, a baseball card, or a cryptocurrency token. That doesn't mean you're doomed. It means your risk is tied to market mood, not just economic output. Mood can flip faster than fundamentals.

Speculation isn't new-what changes is the story we tell ourselves

Speculation has been with us as long as trade. What changes over time is the *justification* that makes speculation feel like investing. Every mania invents a sentence that turns a hot potato into an heirloom.

Sometimes it's technology: "This changes everything." Sometimes it's scarcity: "They're not making any more of it." Sometimes it's democratization: "Regular people finally have access." Sometimes it's inevitability: "It can only go up from here." The sentence varies; the function is constant. It gives people permission to stop asking what the asset produces and start focusing on what the chart suggests.

The market, meanwhile, quietly rewards the transition—at first. When enough people decide the point is resale, the demand curve can steepen. Price rises draw attention; attention draws more buyers; more buyers push price higher. It feels like discovery, as if the market is finally waking up to the asset's "true value." Often, the "truth" is simply that more people showed up.

What makes this psychologically powerful is that it offers two payoffs at once: money and belonging. When prices climb, profit feels like proof of intelligence. When your friends are climbing too, it feels like proof of community. Bubbles are rarely just financial events. They are social events with tickers.

Why money is different from everything else

A chair has a purpose. A loaf of bread has a purpose. Money's purpose is slippery: money is a claim on other people's effort. It is a socially agreed token that can be exchanged for almost anything, which means it's uniquely sensitive to confidence. If you doubt the token, you doubt everything priced in it.

That's why money and near-money instruments—bank deposits, Treasury bills, money market funds—are the hidden foundation under every market story. When people feel safe, they are willing to hold riskier assets because the "escape hatch" feels reliable. When people feel unsafe, they reach for money-like things, even when those things yield almost nothing. This is not irrational. In a world where the future can punch you in the face, optionality is valuable.

The twist is that safety is relative. In some eras, "cash" means a bank deposit. In others, it means a government bond. In extreme cases, it means a foreign currency or gold or simply inventory—anything you can trade when trust fractures. The engine of finance is not just supply and demand for assets. It's supply and demand for *certainty*.

Price discovery: the market as a machine for processing disagreement

"Price discovery" is the polite term for what a market does all day: it takes disagreement and turns it into a number. You think the stock is undervalued. Someone else thinks it's overpriced. Your opinions don't matter until one of you is willing to commit capital at a specific price. Markets force you to stop speaking in adjectives ("cheap," "expensive") and start speaking in verbs ("buy," "sell," "wait").

The practical beauty of this mechanism is that it can be brutally honest. If you are wrong, the market does not negotiate with your ego. If your thesis depends on a buyer who never arrives, the price will not rise out of sympathy. And if you are right, the market may still torture you for months before it agrees. Price discovery is efficient at producing a number, not at protecting anyone's feelings.

The dark side is that markets can also process *noise*. A rumor, a misread data point, a charismatic narrative, a crowded trade—these are also inputs. The machine doesn't separate wisdom from folly. It simply aggregates urgency. That's why the same mechanism that makes markets so useful in normal times can make them so dangerous in extremes: the market is an amplifier. It takes what people are willing to do and scales it up.

The ``greater fool'' isn't always a fool

The phrase "greater fool" is usually delivered with a smirk: you buy something overpriced hoping to sell it to a greater fool later. But the uncomfortable truth is that the "greater fool" can be a rational actor in a system that rewards short time horizons.

If a fund manager is judged quarterly, "intrinsic value" can become a luxury. If your neighbor is making money flipping condos, refusing to participate can feel like refusing to compete. If leverage is cheap and plentiful, a small price move can produce a huge return on your cash—making the temptation almost mathematically irresistible. In those conditions, the greater-fool game can look less like stupidity and more like adaptation.

This is how manias recruit smart people. They don't do it by turning intelligence off. They do it by changing the payoff structure so that intelligence serves the wrong master. The mind starts building arguments for why a rising price is evidence of a rising future, rather than merely evidence of rising demand. The chart becomes a storyteller.

The market, for its part, does not care why you buy. It cares only that you do. That indifference is a feature in normal times; it's a vulnerability in bubbles. A market can float for a long time on the demand created by speculation because speculation itself creates demand. The engine can run on its own exhaust—until it can't.

A simple discipline: separate ``price stories" from ``cash stories"

One way to keep your footing is to force every asset into one of two narratives.

A *cash story* is about what the asset produces: rent from a property, interest from a bond, profits from a business, dividends from a stock. You can argue about the stability of those cash flows, their growth, and the risks around them. But you're anchored to something that exists outside the market's mood.

A *price story* is about who will pay what later. Sometimes that's perfectly legitimate—many trades are about relative pricing, liquidity, and timing. But a price story is inherently reflexive: your expected return depends on other people's expectations. Reflexive systems can behave beautifully for a while and then violently snap.

Neither story guarantees safety. Cash flows can evaporate. Rent can stop. Businesses can fail. Bonds can default. But the act of labeling the story changes your behavior. It reminds you what must be true for you to win. It also tells you what kind of danger you should be watching for: deterioration in economics (cash story) versus deterioration in sentiment and liquidity (price story).

This distinction is also the moral center of many financial panics. When a market is dominated by cash stories, declines are painful but often digestible: people reassess earnings, revise forecasts, lower prices. When a market becomes dominated by price stories, declines can become existential: if the only reason to own is resale, then a falling price is not just a bad outcome—it is the destruction of the asset's purpose.

Actionable takeaway: ask one question before you buy

Before buying any asset, write down—plainly, in a single sentence—which of these is your primary reason:

> "I'm buying this for the cash it will generate," *or* "I'm buying this because I expect I can sell it to someone else for more."

If it's the first, name the cash flow (rent, interest, profits/dividends) and what would have to happen in the real world to threaten it. If it's the second, name the specific condition that would make the buyers disappear (higher rates, tighter credit, a regulatory change, a broken narrative, simple boredom). That one sentence won't prevent losses, but it will prevent self-deception—the most expensive expense in finance, and the one bubbles always charge at the worst possible moment.

1.2 Time and Uncertainty: Unlocking the Math of Valuation

Someone offers you a choice: $1,000 in your hand right now, or $1,000 five years from now. If you hesitate, you're either very polite or very suspicious. The "take it now" instinct isn't greed; it's a recognition that time is not empty space. Time is where layoffs happen, cars break down, opportunities show up unannounced, and inflation quietly chews on yesterday's dollars. Finance begins with that ordinary human preference: money today is more useful than the *same* money later.

That preference is so universal it becomes a kind of gravity. If price is the truce between buyers and sellers, interest rates are the slope of the ground they're standing on. Change the slope and the same crowd behaves differently. When rates are low, the future feels close enough to touch. When rates are high, the future feels far away and expensive. That single shift can lift or crush entire markets without changing a single factory, product, or paycheck.

The time value of money: paying rent to the calendar

The *time value of money* is the idea that a dollar has a "rent"—a cost for keeping it tied up. If you lend someone money, you charge interest because you're giving up flexibility and taking on the risk of not being repaid. If you invest money, you expect a return for the same reasons. The calendar isn't just a way to count days; it's a toll road.

The simplest way to see it is to flip the usual question. People often ask, "How much will my money grow?" Finance often asks the reverse: "How much is that future money worth *today*?" That move—pulling future cash back into the present—is valuation's core trick. It's also why valuation is less like guessing and more like translating: you're converting time into dollars.

Suppose you can earn 5% per year in a safe place. Then $1,000 today becomes $1,050 in a year. Turn that around: $1,050 received in a year is worth $1,000 today *if* 5% is the right rate for the job. The math is simple. The implications are not. Once you accept that the future must be "discounted," you've accepted that changing the discount rate changes almost everything people call "value."

Discount rates: the lever hiding under every price

A *discount rate* is the rate you use to translate future cash into today's dollars. In plain English, it's the return you demand for waiting and for taking risk. Higher discount rate means you're less willing to pay up today for a promise tomorrow. Lower discount rate means you're more willing.

This is where interest rates become market-wide gravity. Even when you're valuing a business, you're still in conversation with the risk-free rate in the background, because that rate is the baseline alternative: what you can earn without taking much credit risk. When the baseline rises, everything that competes with it has to justify itself again. You can still buy the dream of far-off profits, but the dream gets more expensive to hold.

Here's the brutal piece: the farther away the cash flows are, the more sensitive the price is to the discount rate. That's why long-dated bonds can swing hard when rates move, and why "growth" stocks (whose hoped-for cash flows are often years out) can feel like they get hit twice when rates rise: once in the spreadsheet, and once in the mood.

Immediately after that explanation, it helps to see the shape of the relationship:

The curve looks almost unfair. At 0%, $100 ten years from now is worth $100 today. At 10%, it's worth less than $40. Nothing "happened" to the $100. Only the rent charged by time changed. If you ever wonder why markets can reprice so violently when central banks move rates, the answer often isn't in the headlines—it's in that curve.

Valuation as disciplined time-travel

Valuation is just time-travel with rules. You list the cash you expect to receive and when you expect to receive it. Then you discount each piece back to today using a rate that reflects waiting and risk. Add them up and you get a present value.

For a bond, the cash flows are explicit: coupon payments and principal repayment. For a rental property, they're rents minus expenses, with a sale price at the end. For a business, they're the cash left over after paying everyone else—the money that can be reinvested or returned to owners.

The reason this feels harder than it sounds is that most assets aren't honest like bonds. A business doesn't hand you a schedule of future profits. A property doesn't guarantee tenants. Even a government bond carries inflation risk: you might get paid, but in weaker dollars. Valuation forces you to admit what you're guessing about and what you're not.

A useful mental shift: a price is not only a number you observe; it's also a set of beliefs hiding in a trench coat. If you know the price and you have a model of the cash flows, you can work backward to infer what the market must be assuming—about growth, margins, default risk, or how long good times last. That's why professionals talk about "implied" growth and "implied" returns. A market price is a story with math underneath it.

Net present value: the bouncer at the club

The *net present value* (NPV) rule is valuation's most practical weapon because it's so blunt it's hard to charm. NPV says: discount every cash inflow and outflow back to today, add them up, and accept the deal only if the total is positive.

If that sounds cold, it is—and that's the point. Finance is full of opportunities that feel exciting because they're large, complex, and flattering. NPV doesn't care. It's the bouncer at the door who looks at your ID, not your personality. A positive NPV means the investment, at your chosen discount rate, is expected to create value beyond what you require for waiting and risk. A negative NPV means you're paying too much for what you're likely to get.

NPV also reveals a subtle trap: the biggest disagreements are often not about the cash flows, but about the discount rate. Two people can forecast similar rents or profits and still come to opposite conclusions because one demands a higher return than the other. The optimist isn't always "better at forecasting"; sometimes the optimist is simply charging the future less rent.

If you want a quick reality check when you're tempted by an investment pitch, ask: "What discount rate makes this look attractive?" If the answer is "a rate that assumes nothing ever goes wrong," you've found the hidden hinge.

No-arbitrage: why identical cash should have identical prices

Markets have a strange kind of internal policing. When two different-looking assets deliver the same cash flows in the same states of the world, they shouldn't have meaningfully different prices. If they did, traders would buy the cheaper one, sell the more expensive one, and lock in a near risk-free profit. That pressure tends to erase blatant inconsistencies.

This idea is called the *no-arbitrage*. It sounds academic, but it's really the market's version of "you can't get free money forever." The concept matters for a deeper reason: it's what makes valuation more than opinion. If prices weren't constrained by consistency, every asset would be pure storytelling. No-arbitrage forces stories to reconcile with math, at least roughly.

It also explains why bubbles often show up first in the seams—places where comparables stop lining up, where one slice of a market becomes detached from another, where the same cash flow is being priced differently depending on packaging. In the mid-2000s, for example, complicated credit structures created exactly the kind of packaging that can blur comparability. When the packaging works, it looks like innovation. When it doesn't, it looks like a magic trick you can't repeat.

Risk: when you know the game and you can price the odds

People use "risk" to mean "bad things might happen," but in finance *risk* has a more specific flavor: it's uncertainty you can put numbers on. Roulette is risky, but the odds are known. Insurance works because car accidents are uncertain for any individual driver but statistically regular across millions of drivers. With enough data and stability, you can price risk.

In markets, measurable risk shows up in things like volatility (how much prices bounce around), default rates (how often borrowers fail to pay), and correlations (whether assets tend to fall together). These are imperfect, but they're at least something you can estimate, stress, and argue about.

Measured risk is why finance has models, and why those models sometimes help. If you're lending to thousands of borrowers with similar characteristics, historical patterns can be informative. If you're pricing a bond with clear legal terms, you can do real analysis. If you're building a diversified portfolio, you can make sensible decisions about how much bumpiness you can tolerate.

But measurable risk has a dirty secret: it can make people overconfident, especially when the recent past has been calm. A model fed on tranquil data will politely suggest the world is tranquil. It's not lying. It's just trapped in its own sample.

Uncertainty: when you don't even know what could happen

Then there's *uncertainty*—the kind that doesn't fit neatly in a spreadsheet because the distribution isn't stable and the rules may be changing. Uncertainty is not "a higher level of risk." It's a different species.

Uncertainty is waking up to a headline you didn't know could exist. A bank run in an era that "doesn't do bank runs anymore." A sudden ban, a sudden war, a sudden pandemic. A technology that makes yesterday's business model look like a horse-drawn carriage parked next to a rocket. Under uncertainty, the problem isn't that you don't know the odds; it's that you don't know the list of outcomes.

This is where panic is born. Panic isn't merely fear of losses; it's fear of the unknown *shape* of losses. When uncertainty surges, investors stop arguing about the exact discount rate and start questioning the cash flows themselves. "Will I be paid?" turns into "Will this even function?" Liquidity dries up because nobody wants to be the person who bought right before the rules changed. Prices can gap down not because everyone agreed the asset is worth less, but because fewer people are willing to name a price at all.

Uncertainty also has a cruel timing feature: it tends to appear exactly when leverage is high and confidence has been mistaken for stability. In booms, people don't just buy assets; they buy narratives about why the future is manageable. Uncertainty is the moment those narratives get punctured.

Why low rates can feel like a lullaby---and why that matters for manias

When discount rates are low, the present value of distant cash flows rises. That's a mechanical fact. But it also changes culture. Low rates don't just change spreadsheets; they change what people are willing to *believe*. If safe returns are tiny, the temptation to reach for stories with bigger payoffs grows stronger. The financial world starts to treat patience like a form of courage: "Hold the long-duration asset; the future will reward you." Sometimes it does. Sometimes the future sends a different email.

This is one reason bubbles can be hardest to spot in the moment. The math is pointing in the same direction as the mood. Rising prices feel justified because a lower discount rate really does raise present values. That legitimate effect can blend seamlessly into illegitimate extrapolation: from "future cash is worth more today" to "the future will be amazing and uninterrupted." The first is valuation. The second is mythology.

When rates rise, the reverse happens. It's not just that discounted cash flows shrink. The cultural permission to be optimistic gets revoked. Deals that depended on cheap money start to look like they were built on air. The market's disagreement shifts from "how big will the upside be?" to "who is forced to sell?"

Actionable takeaway: run a two-rate valuation before you trust any price

Before buying an asset because it "looks cheap" or because a chart has been kind to it, do a simple exercise:

Write down your expected cash flows (even rough ones) and calculate the present value twice—once with a "comfortable" discount rate and once with a "stress" discount rate that reflects a harsher world (higher rates, lower growth, more friction). If you don't want to build a full model, even a back-of-the-envelope version works: the goal is not precision; it's *sensitivity*.

If the investment only looks attractive under the comfortable rate, you're not buying an asset—you're buying a particular interest-rate environment and a particular mood. If it still looks reasonable under the stress rate, you've found something sturdier: a deal that can survive the calendar's rent going up and the world getting weird, which is exactly when true uncertainty tends to arrive.

1.3 The Plumbing of Panic: Liquidity and Infrastructure

On a normal day, selling a stock feels like turning a doorknob. Tap a button, watch a confirmation flash, and move on with your life. The ease is so smooth it's almost insulting: *how can something as serious as money move with the friction of a text message?* That sensation is the financial equivalent of believing the lights turn on because your finger flipped a switch. The truth is less magical and more fragile. There's a whole grid behind the wall, and it only looks like magic when nobody is stressed.

Liquidity is the word people reach for when that magic starts to fail. It's often treated as a synonym for "cash," but cash is only one form of liquidity-like water is only one form of plumbing. Liquidity is the ability to transact *when you want to* without taking a humiliating haircut. It's the difference between selling a thing and begging someone to take it. It's also the difference between a small market wobble and a full-body financial seizure.

A useful way to think about liquidity is as a promise the system makes to you: *if you need to change your mind, you can.* Markets are built on that promise. When you buy a share, you're not committing to own it forever; you're buying an asset that comes with an exit door. In calm times, the door is wide and well-lit. In panics, everyone discovers the door was always a single-file staircase.

Three liquidities: the asset, the market, and the firm

People say "liquid" like it's one thing. It isn't. There are at least three different meanings, and confusing them is how smart people sound foolish in crises.

An *asset* is liquid when it can be sold quickly with little loss in value. U.S. Treasury bills are the classic example: they trade in enormous size with tiny price impact. A rare painting is illiquid: even if it's valuable, you may need weeks of marketing, negotiation, and luck.

A *market* is liquid when it can absorb buying and selling without lurching. This is about depth—how many orders are waiting at nearby prices—and resilience—how quickly new orders appear after a shock. A market can look liquid when everyone is cheerful and then suddenly reveal it was thin ice all along.

A *firm* is liquid when it can meet its short-term obligations. This is about cash and funding, not about profitability. A company can be solvent (assets exceed liabilities) and still die from illiquidity if it cannot make payroll or roll over short-term debt. That's not a metaphor; it's a common cause of financial accidents.

These distinctions matter because panics are often not fundamentally about *value*. They're about *timing*. A solvent firm can't wait for its assets to "recover" if the bill is due today. A good asset can become a bad asset if you're forced to sell it on the worst day of the year.

The hidden cast: brokers, exchanges, clearinghouses, and settlement

The trading app is the stage; the plumbing is backstage. When you place an order, your broker routes it to a trading venue-an exchange or another marketplace-where it can match with someone else's order. That match produces a trade: two parties now owe each other. One owes cash; the other owes shares.

But a trade is not the same thing as a completed exchange. Between "we agreed" and "it's final" sits a short, crucial stretch of road called *clearing and settlement*. Clearing determines who owes what after netting buys and sells together. Settlement is the actual delivery: cash moves one way, securities the other, and the deal becomes real.

In everyday life, you settle instantly. You hand someone $20 and receive a book. Financial markets often don't work that way because the volumes are enormous and the participants trade constantly. Instead, the system uses infrastructure to keep track of obligations, reduce complexity, and lower the risk that one failure becomes everybody's problem. That infrastructure is why you can trade at all with people you've never met and institutions you don't trust personally.

A *clearinghouse* sits in the middle of many markets like a stern accountant with a shock collar. It becomes the buyer to every seller and the seller to every buyer, guaranteeing performance—*as long as* participants post enough collateral and the rules hold. This centralization is a blessing and a pressure point. It reduces the mess of bilateral promises, but it also concentrates risk in a single institution whose job is to remain boring during the most exciting moments in finance.

Netting: the quiet superpower that shrinks obligations

If you imagine every trade settling individually, the system looks absurd. A large broker might buy and sell the same stock thousands of times a day. Settling each trade gross would require enormous transfers of cash and securities—like paying for every grocery item separately with a wire transfer.

Instead, modern systems rely on *netting*: adding up what you owe and what you're owed, then settling the difference. Netting is why the market can handle oceanic trading volume without requiring an ocean of money to slosh around every hour.

The difference is not subtle. On a typical day, multilateral netting can compress the overwhelming bulk of obligations into a small net amount—think on the order of roughly 98–99% reduction from gross to net. That number sounds like a magic trick, but it's just arithmetic applied at scale. It is also why the system can look incredibly robust right up until it doesn't. When netting works, it saves liquidity. When netting is disrupted—or when participants can't fund even the smaller net obligations—the stress shows up all at once.

Netting also changes the nature of panic. In a gross-settlement fantasy, you'd see the strain building trade by trade. In the real world, netting can hide strain until the moment funding fails, at which point the problem isn't gradual. It's cliff-like.

Settlement speed: safer isn't always simpler

"Why not just settle instantly?" sounds like a child's question until you realize it is also a regulator's question, a technologist's question, and a trader's question—often asked with different motives. Faster settlement reduces the time you're exposed to a counterparty failing between trade and delivery. That's good.

But there's a tradeoff that feels counterintuitive: faster settlement can *increase* the system's demand for intraday liquidity. If you lose the ability to net obligations over a day (or over multiple trades), you may need to fund many more gross payments. In other words, you reduce credit exposure but require more cash-on-hand at the exact moments stress is highest.

Finance is full of these conservation laws. Risk doesn't disappear; it changes form and location. Speed can reduce one kind of vulnerability while intensifying another. The plumbing choices—netting rules, margin rules, settlement timing—shape how a panic travels through the system, and where it finds weak joints.

Finality: the difference between a promise and a fact

In a crisis, the most precious commodity is not return. It's certainty. The system is held together by the belief that payments and deliveries actually *finish*. That's why payment systems that settle in central bank money with immediate finality are so important. When settlement is final, the transaction is done; it can't be unwound because someone regrets it or because a chain of promises snapped elsewhere.

Finality sounds legalistic until you imagine the alternative: a world where you sell an asset, believe you have the cash, spend it, and then discover the payment was reversed because the other party failed. That's not just inconvenient; it's destabilizing. It forces everyone to hoard, delay, and distrust—exactly the behaviors that suffocate liquidity.

To keep the pipes flowing during the day, some systems provide *intraday* credit (often called daylight credit), with controls like caps and collateral requirements. The point is not to encourage recklessness; it's to prevent a traffic jam from becoming a pileup. A modern financial system cannot run if every participant insists on being fully pre-funded every minute of the day. That would be like requiring every driver to own the entire highway before merging.

Market liquidity vs. funding liquidity: the two blades of the same scissors

There's another distinction that becomes life-or-death in panics: *market liquidity* versus *funding liquidity*.

Market liquidity is the ability to sell an asset without moving the price too much. Funding liquidity is the ability to borrow or raise cash against assets—through repo markets, credit lines, margin loans, and short-term funding. In quiet times, these two reinforce each other. When you can borrow easily, you can hold inventory of assets, make markets, and provide bids. When markets are liquid, lenders feel safer accepting assets as collateral.

In stress, the relationship can invert and become vicious. Lenders raise margins and haircuts (the discount they apply to collateral). Dealers and funds suddenly need more cash to hold the same positions. To raise that cash, they sell assets. Those sales push prices down and widen spreads, worsening market liquidity. The worsening market liquidity makes lenders even more nervous, leading to higher haircuts and more margin calls. That's a liquidity spiral: not a single bad decision, but a feedback loop that punishes everyone simultaneously.

This is one reason panics feel like physics rather than psychology. Individuals can be calm and still be forced to act by the mechanics of funding and collateral. A margin call does not care whether you "still believe in the long-term." It cares whether you have cash by 2 p.m.

Why ``there will always be a buyer" is a fair-weather myth

In boom times, it's easy to mistake activity for depth. Lots of trading creates the impression that there are always counterparties. But much of that activity is conditional. Market makers quote prices because they expect to manage risk continuously. Funds buy because they expect they can sell. Banks lend because they expect collateral will remain liquid. The system is a chain of "I will, as long as."

When confidence cracks, those conditions change. Market makers widen spreads or step back. Funds hit redemptions and must sell. Lenders demand more collateral or refuse to roll funding. The "buyer" you counted on is suddenly in the same hallway as you, heading for the exit.

This is also where the earlier idea of discount rates meets plumbing. When uncertainty spikes, required returns rise, which means discount rates rise, which means prices fall. Falling prices then feed back into collateral values and margin requirements, which tightens funding liquidity, which forces more selling, which pushes prices down further. The math and the plumbing can align in the worst possible direction.

Clearinghouses and margin: shock absorbers that can stiffen in a crash

Clearinghouses reduce counterparty risk by requiring participants to post margin—collateral that covers potential losses if a participant defaults. In normal times, margin is a seatbelt. In panic, margin can become an accelerator if it rises sharply.

This is not because clearinghouses are malicious; it's because they are designed to survive. If volatility rises, the potential loss from a default rises, so margin models demand more collateral. Participants scramble to find cash or high-quality collateral. To do that, they sell what they can. That selling increases volatility. The model responds. The system tightens. Everyone feels the belt pulling harder.

A well-designed clearing system is still vastly better than a world of uncollateralized promises. But the tradeoff is real: the safety mechanisms often demand liquidity at precisely the moment liquidity is scarce. The system protects itself by pulling resources forward in time. During stress, time itself becomes expensive—again.

A panic is often a settlement problem wearing a price chart

When people talk about market crashes, they often describe them as collective insanity: fear, greed, herd behavior. Those are real. But panics also have an operational spine. If settlement fails, if funding dries up, if collateral calls cascade, the market can move regardless of what anyone "believes."

That's why some of the most terrifying moments in financial history have a specific feel: not just prices falling, but *transactions* failing. Trades don't clear. Quotes disappear. Lenders stop answering. The world shrinks to a handful of acceptable collateral types and a single priority: survive today. In those moments, "value" becomes a luxury word. Liquidity becomes the only word.

And once you understand that, a lot of baffling behavior stops being baffling. Why would someone sell a good asset at a bad price? Because it's liquid and they need cash now. Why would someone hoard cash earning little? Because the option to act later is priceless when the rules feel unstable. Why do correlations jump toward one in crises? Because forced selling is not picky; it sells what can be sold.

Actionable takeaway: stress-test your exit, not just your entry

Before you buy any asset—or before you lever one you already own—ask a question that feels pessimistic but pays for itself in calm times:

> *"If I needed to exit quickly during a bad week, what would actually happen?"*

Answer it concretely. Where would you sell (which venue or broker)? What order type would you use? How wide might spreads get? Would you face margin calls or collateral haircuts? Would your funding roll, or could it vanish overnight? If the honest answer is "I'm not sure," treat that uncertainty as part of the price you're paying.

Liquidity is like oxygen: you only notice it when it's missing. The practical discipline is to notice it *before* you need it—while the door is still wide, the lights are still on, and the engine still sounds smooth.

1.4 The Amplifier: Credit, Money Creation, and Debt

A strange thing happens the first time you buy something expensive with borrowed money: you feel richer without getting richer. The car is in the driveway, the house keys are in your hand, the stock position is in your account—and yet the money was never really *yours*. Credit is the financial world's most intoxicating illusion: it lets you consume or invest today as if tomorrow has already arrived, then sends you the bill slowly enough that you can forget the trick.

That trick is not a side show. Credit is the amplifier bolted onto every boom and every bust. When it's abundant and cheap, it does more than increase spending; it changes the kinds of bets people are willing to place, and the prices they're willing to pay. When it tightens, it does more than slow things down; it forces liquidations, triggers margin calls, and turns "paper losses" into real ones. If prices are a crowd's opinion and liquidity is the ability to act on that opinion, credit is the force that makes the crowd loud.

Credit: not money, but the right to spend like it is

Credit is permission. It's an agreement that you can take control of resources now and settle up later. In everyday life, credit shows up as a mortgage, a credit card, a car loan, a student loan. In markets, it shows up as margin loans, repo funding, corporate bonds, bank credit lines, and a thousand other pipes carrying the same substance: time.

The emotional appeal is obvious. Credit smooths life. It lets families buy a home without waiting decades. It lets businesses build factories before they've accumulated the cash. It lets governments fund projects whose payoff spans generations. Used well, credit is civilization's way of building ahead of its savings.

Used badly, credit turns the future into collateral for present confidence. And confidence, as markets repeatedly demonstrate, is not a stable material.

Where ``money" actually comes from (most of the time)

Many people carry a simple story in their heads: banks take deposits from savers and lend them out to borrowers, like a careful librarian loaning books. It's intuitive, tidy, and incomplete.

In modern economies, most of what people use as money is not paper currency; it's bank deposits—the numbers in your checking account. And those deposits are largely created when banks make loans. When a bank approves a mortgage, it doesn't typically rummage around for someone else's stored pile of dollars and hand them over. It credits the borrower's account with a deposit and records a loan asset on its own books. In one stroke, the bank has created a new deposit—new purchasing power—alongside a new debt.

This is not a conspiracy and not a loophole. It's the design. Banking is a business of balance sheets, not of warehouses.

That design is why credit can expand quickly. It is also why credit can contract quickly. When loans are repaid or written off, deposits can shrink. When banks tighten standards, new deposit creation slows. The money supply many people *feel* in the economy—how easy it is to borrow, how willing lenders are—can change faster than the physical economy underneath it.

So what stops banks from creating infinite money?

If banks can create deposits by lending, the next question is uncomfortable: why not lend endlessly?

The constraint is not a simple "reserve multiplier" where banks lend out a fixed fraction of deposits like a mechanical pump. In practice, the binding constraints are messier and more human: profitability, solvency, regulation, and risk management. Banks must believe they will be repaid. They must hold enough capital to absorb losses. They must satisfy liquidity rules and funding realities. They must manage reputational and legal risk. They must survive.

Central banks influence the entire process by setting the price of short-term money—interest rates—and by shaping the environment in which banks fund themselves. Higher policy rates generally raise borrowing costs, cool demand for loans, and can tighten financial conditions. Lower rates tend to do the opposite. Sometimes central banks also buy assets (quantitative easing) or use other tools that affect liquidity and market functioning. But the crucial point is this: the "creation" of money through lending is neither unlimited nor purely mechanical. It is pro-cyclical. It expands when the world feels safe and contracts when the world feels dangerous—exactly when the economy would prefer the opposite.

That pro-cyclicality is the amplifier in its purest form.

Debt as a growth machine---and a fragility machine

Debt is often painted as moral failure. That's comforting because it makes the world simple: good people save, bad people borrow, and crises happen when the bad win. Reality is less satisfying and more useful.

Debt is a tool for matching long-lived projects with immediate resources. A bridge that lasts fifty years is not easily funded by waiting for spare change to accumulate. A company building a network, a factory, or a fleet doesn't want to wait until retained earnings have piled up. Credit pulls investment forward.

But the same tool creates fragility because it introduces fixed obligations into a world of variable outcomes. A business with no debt can have a bad year and limp through. A business with heavy debt can have a bad quarter and die. A household can survive job instability if it has flexibility. A household with a large mortgage and thin savings can be pushed into default by one unlucky month.

Debt is not inherently evil; it is inherently *nonlinear*. Small changes in income or asset prices can cause disproportionately large changes in survival prospects.

Leverage: buying a giant shadow with a small flashlight

Leverage is the specific form of credit that turns that nonlinearity into a weapon. *Leverage* means controlling a large exposure with a relatively small amount of your own money by borrowing the rest. It can make you look like a genius in a rising market and erase you in a falling one, sometimes faster than your brain can process what happened.

Here is the cleanest way to see it. Suppose you have $10,000 and you buy $10,000 of an asset. If the asset rises 10%, you make $1,000: a 10% return.

Now suppose you have $10,000, but you borrow $90,000 and buy $100,000 of the same asset. If it rises 10%, you make $10,000—roughly doubling your money (ignoring interest costs). You didn't become smarter. You turned up the volume.

Turn the market the other way and the math becomes cruel. If the asset falls 10%, you lose $10,000—your entire equity is gone. The asset didn't need to collapse. It only needed to wobble.

That's leverage: the amplifier that makes ordinary volatility feel like fate.

A little movement on a large position
can crush a small equity cushion.

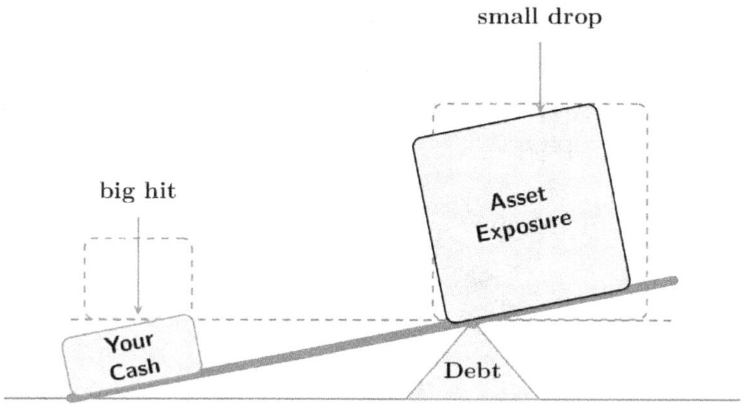

The diagram is cartoon-simple because leverage itself is cartoon-simple. Complexity arrives later, when leverage is layered: a mortgage inside a bank balance sheet inside a securitization inside a leveraged fund funded overnight. Each layer feels like someone else's problem until the chain tightens and everyone discovers they were holding the same rope.

Margin: the silent clause that turns a loss into a liquidation

Leverage is rarely "set it and forget it." It comes with conditions—especially in markets. If you borrow to buy securities, your broker requires *margin*: your own money that sits as a buffer against losses. If prices fall far enough, your buffer shrinks and you may face a *margin call*: deposit more cash or collateral, or your position will be sold out from under you.

This is where the amplifier becomes a weapon aimed at its owner. A leveraged investor is not just exposed to price moves; they are exposed to the *timing* of price moves. Even if the asset recovers later, they may not survive the path. That's why leveraged markets can crash faster than fundamentals deteriorate. The selling is not always a change of mind. Sometimes it's a contract enforcing reality.

Margin calls are also contagious. Forced selling pushes prices down, which triggers margin calls for others, which forces more selling. When people describe a crash as "everyone rushing for the exit," margin is often the person at the door pushing them through.

The credit cycle: optimism manufactures balance sheets

Credit is not a constant; it breathes. In expansions, defaults feel rare, collateral feels solid, volatility feels tame, and lenders become generous. Standards loosen gradually, then suddenly. Borrowers who used to be rejected become "underserved markets." Deals that used to look reckless become "innovative." The economy can genuinely improve during this phase, which makes it even harder to see the danger. Rising incomes and rising asset prices validate the lending that helped cause them.

But the same process accumulates fragility. More debt means more fixed obligations. Higher asset prices mean thinner cushions. Business models appear that only work under easy funding. A whole ecosystem can form around the assumption that refinancing will always be possible—an assumption that holds until the day it doesn't.

When conditions tighten—because rates rise, because losses appear, because a shock hits—credit stops being accommodating. Lenders demand more compensation or more collateral. Borrowers who relied on rolling short-term funding find the door closed. The cycle turns from expansion to contraction, and because money creation is tied to lending, the contraction can feel like the economy's bloodstream pulling back from its extremities.

Economists and regulators watch indicators like credit growth relative to the economy—credit-to-GDP measures—because rapid credit expansion has a habit of preceding systemic trouble. It's not fortune telling; it's pattern recognition. When credit grows much faster than the economy's ability to service it, the system is quietly loading the spring.

Minsky's uncomfortable idea: stability breeds instability

The late economist Hyman Minsky offered a line that sounds like a paradox until you've lived through a boom: stability encourages risk-taking, and risk-taking eventually destroys stability. When things go well for long enough, people stop pricing disaster. They don't do it because they're stupid; they do it because *recent experience* becomes their evidence.

Under prolonged calm, financing shifts. Borrowers move from "I can pay interest and principal from my income" to "I can pay interest, and I'll refinance the principal later" to "I can't even cover interest unless prices keep rising." These aren't just categories in a paper; they're habits you can watch in real time: teaser rates, interest-only loans, covenant-light corporate debt, funding structures that require perpetual market access.

The scary part is that the system can look healthiest near the end of the boom. Defaults are low. Asset prices are high. Volatility is low. Confidence is high. The amplifier is humming smoothly. That's exactly when leverage is easiest to justify—because the recent past has been generous enough to sponsor the argument.

Then a shock arrives. Sometimes it's big, sometimes it's embarrassingly small. The system doesn't break because the shock is unprecedented; it breaks because the balance sheets were built with no tolerance for surprise.

Why booms feel like merit---and busts feel like betrayal

Credit-driven booms are emotionally persuasive because they reward behavior quickly. If you lever up in a rising market, you get paid sooner and louder than the cautious person. That feedback doesn't just change portfolios; it changes social status. It changes which opinions sound "smart." It changes what people are willing to say out loud at dinner parties.

Busts feel like betrayal for the same reason. The rules didn't feel like rules when they were working. They felt like the natural order. When credit tightens, the same people who praised discipline begin demanding rescue. That's not hypocrisy so much as the psychological whiplash of discovering you were living in a subsidized reality.

Understanding credit as an amplifier helps cut through the moralizing. The question is not "Were people greedy?" The question is "How much leverage did the system allow, and how quickly did that leverage become self-reinforcing?" Greed is timeless. Leverage is a design choice, a regulatory choice, a market-structure choice, and a human choice made under the influence of recent returns.

Actionable takeaway: measure leverage in plain English before you admire a return

Before you copy anyone's performance—your own past self included—translate the strategy into one sentence of leverage math:

> *"If the asset falls X%, what happens to my equity, and will I be forced to sell?"*

Pick a real X, not a polite one. If a 10% drop wipes you out, you are not investing—you are balancing on a knife edge and calling it confidence. If a 20% drop would trigger margin calls, you are not just betting on direction; you are betting on a smooth path.

Credit is powerful because it can build real prosperity. It is dangerous because it can turn normal uncertainty into a solvency event. The habit that keeps you alive is simple: never let borrowed money convince you that volatility is smaller than it is.

Chapter 2

Valuation, Narratives, and the Madness of Crowds

Sir Isaac Newton, after losing a fortune in the South Sea Bubble, famously lamented that he could calculate the motions of heavenly bodies, but not the madness of people. Why do rational people make irrational financial decisions? This chapter steps away from the spreadsheets to explore the human element of finance. We will examine how compelling stories trump boring math, why our brains are hardwired to herd, and why professional money managers are often forced to dance while the music plays, even when they know the floor is about to collapse.

2.1 Anchors of Value: Fundamentals versus Stories

A bubble rarely starts as a lie. It usually starts as a truth with good lighting.

Someone points at a real change—a new technology, a new market, a new policy regime, a new source of easy credit—and says, "This time the old rules don't fit." At first, that's not crazy. When railroads knitted together the United States, when radio turned air into a business, when the internet made distribution nearly free, the world genuinely did change. The trouble begins when "the old rules don't fit" quietly mutates into "there are no rules." That's the moment prices stop being estimates and start being votes.

A useful way to think about investing is to imagine two forces tugging on every asset. One is gravity: the boring arithmetic of cash flows, risk, and time. The other is weather: stories, moods, and contagious optimism. Gravity doesn't make headlines. Weather does. Most of the time, gravity wins—but the weather can throw a spectacular party before the cleanup.

What ``value'' actually means when nobody's watching

An asset—a stock, a bond, a rental property, a business—is worth the cash it can produce for its owners, adjusted for two inconveniences: (1) those cash flows arrive in the future, and (2) the future is unreliable.

That's it. Everything else is a shortcut, a proxy, or a sales pitch.

If you buy a share of stock, you are buying a claim (a small one, but real) on what the business can eventually send back to you—dividends, buybacks, or, in the long run, the option to sell your share to someone else who expects dividends and buybacks. If a business never returns cash to owners and never plausibly could, the share certificate is not a financial asset so much as a collectible. Collectibles can soar, but they do not have an intrinsic engine. They have a crowd.

This is where the math earns its keep. It gives you a way to translate a company's messy, story-filled future into a single number you can argue about. Not a perfect number. Not a sacred number. Just a disciplined estimate with assumptions you can list out without blushing.

Valuation multiples: quick-and-dirty anchors (and how they get cut loose)

Most investors don't start with a full-blown valuation model. They start with a multiple—a ratio that compares the price to some measure of fundamentals.

The most famous is the Price-to-Earnings ratio (P/E):

$$P/E = \frac{\text{Price per Share}}{\text{Earnings per Share}}$$

In plain English, P/E asks: "How many dollars am I paying for each dollar the company earned?" A P/E of 20 implies you're paying $20 for $1 of annual earnings. That does *not* mean you'll get your money back in 20 years (earnings aren't dividends, and earnings can grow or shrink). But it does create an anchor. It forces a conversation about how good the business must become to justify what you're paying.

Multiples come in a family: Price-to-Sales (P/S), Enterprise Value-to-EBITDA (EV/EBITDA), Price-to-Book (P/B). Each is a different flashlight pointed at the same cave. Sales are harder to fake than earnings but easier to grow unprofitably. EBITDA can be useful but conveniently ignores real costs like interest, taxes, and the fact that machines wear out. Book value can matter for banks and insurers, and can be nearly meaningless for a software company whose most valuable assets go home at 6 p.m.

Here is the trap: multiples feel objective because they look like math. But a multiple is a compressed story. A high P/E is the market saying, "We believe earnings will grow rapidly and reliably, and we're not too scared about the risks." A low P/E is the market saying, "We don't trust those earnings, or we don't think they'll last, or we're nervous."

In a bubble, multiples don't merely stretch. They *change what they are attached to*. Investors migrate from P/E to "P/S is the new P/E," from "profits don't matter yet" to "profits are a choice," from "cash flow later" to "dominance now." When you hear a metric being used not as a tool but as a permission slip, you're watching the anchor line fray.

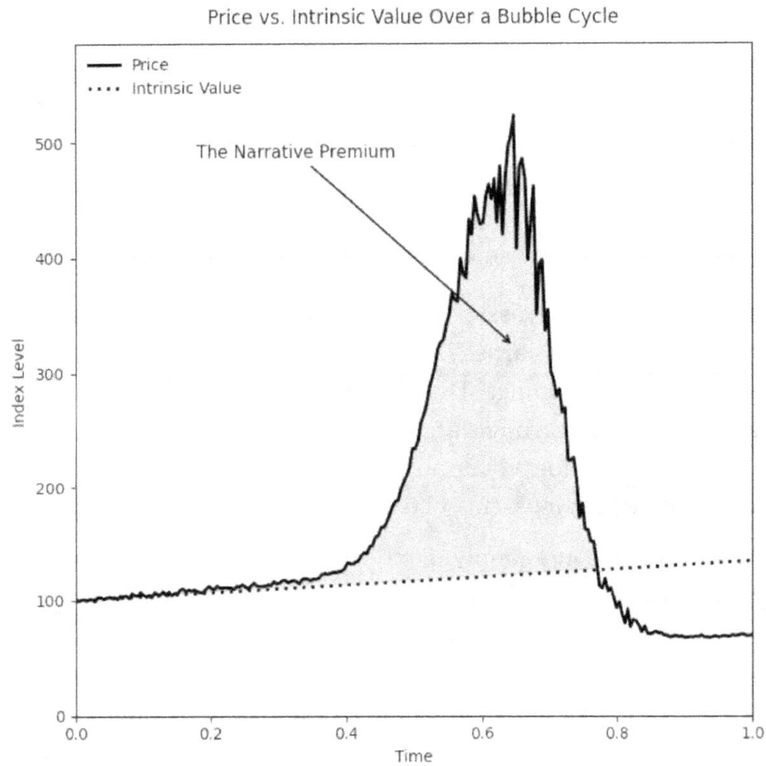

That shaded gap—the distance between what the asset is plausibly worth and what people are paying—is where bubbles live. It's not always pure delusion. Often it's a premium for a possibility: a belief that the company will conquer its market, that a technology will rewrite profit pools, that a new class of buyer has arrived, that central banks will keep the music playing. Possibilities deserve *some* premium. Bubbles happen when the premium becomes the product.

Discounted Cash Flow: turning a story into a spreadsheet that can be questioned

If multiples are a snapshot, Discounted Cash Flow (DCF) is a movie.

DCF starts with a simple idea: money in the future is worth less than money today, because you could invest today's money elsewhere, and because the future can disappoint you. So you forecast the cash the business can generate and discount it back to the present.

A simplified DCF looks like this:

$$\text{Value} = \sum_{t=1}^{N} \frac{\text{CashFlow}_t}{(1+r)^t} + \frac{\text{TerminalValue}}{(1+r)^N}$$

Here, r is the discount rate—the return you demand to compensate for risk and for tying up your money.

The phrase "forecast the cash" is where the drama enters. Forecasts are not facts; they are beliefs with decimals. That's precisely why DCF is valuable. It forces you to admit what you are assuming.

When someone says, "This company deserves a $200 billion valuation," DCF asks, calmly and without sarcasm: "Fine. Tell me what has to be true." What revenue? What profit margins? How much reinvestment to grow? How long does extraordinary growth last before competition shows up with teeth? What happens in a recession? What happens if interest rates rise and the discount rate goes up?

The most dangerous thing about a bubble isn't that people stop using DCF. It's that they use it as a costume. You can make a DCF say almost anything if you crank the assumptions: stretch high growth further into the future, assume margins that would make a monopoly blush, and choose a discount rate that treats risk like an old superstition. The spreadsheet will produce a number with two decimal places and the emotional authority of a lab result. That's not rigor. That's theater.

The ``New Era" narrative: how stories lift prices off the ground

A market is a social place. Prices are formed in public. That means stories aren't optional; they're the coordination mechanism.

In ordinary times, narratives help investors compress complexity. "This is a steady utility." "That is a cyclical automaker." "This is a bank; watch credit losses." These are stereotypes, but they are functional. They keep you from drowning in details.

In bubble times, narratives do something more intoxicating: they grant permission to ignore inconvenient constraints. A "New Era" story has a recognizable rhythm:

- A real innovation appears.
- Early winners are undeniable.
- The story expands from "some companies will benefit" to "all companies touched by this will benefit."
- Valuation shifts from "how much cash will it produce?" to "how big could it get?"
- Skepticism is rebranded as ignorance or bitterness.

The late 1990s offered a masterclass. The internet truly did change distribution, marketing, and commerce. But the market began to price not just changed business models, but changed economics—as if competition had been repealed. "First mover" became "permanent winner." "Eyeballs" became "profits by destiny." When the crowd is euphoric, even a terrible business plan can sound like a land grab.

The housing boom in the mid-2000s had a different flavor of New Era: not "technology rewrites everything," but "housing can't fall nationally," "risk has been dispersed," and "credit innovation has made the system safer." The story didn't need everyone to become a real-estate speculator; it only needed enough people to believe that the floor was solid. Once that belief spread, leverage did the rest.

Crypto cycles have shown a modern version: "code replaces institutions," "scarcity guarantees value," "adoption is inevitable," "we are early," "the old metrics don't apply." Sometimes there is a legitimate technological breakthrough in the background. That doesn't prevent prices from sprinting far ahead of what the underlying use cases can justify in the near term.

What these episodes share is not stupidity. It's the human urge to convert uncertainty into a plotline. A good narrative doesn't merely explain; it recruits. It gives people a role: visionary, early adopter, builder, believer. It offers moral clarity: the future versus the dinosaurs. And it provides a convenient villain: the skeptical analyst with the calculator.

The narrative premium: when the crowd pays extra for a feeling

The narrative premium is the amount investors pay above a sober estimate of value because the story feels powerful, inevitable, and socially validated.

It's easiest to spot when you watch the language around valuation degrade.

Early-stage optimism sounds like: "The upside is meaningful if execution goes well."

Late-stage narrative premium sounds like: "Execution is basically guaranteed because the trend is unstoppable."

You can hear the moment when probability gets replaced by prophecy. Forecasts turn into destinies. Risks become footnotes. Competition becomes a rounding error. And the most telling shift: time stops costing anything. The farther into the future the payoff lies, the more the story is asked to carry—and the less anyone wants to talk about discount rates, dilution, or the fact that customers are fickle.

A practical way to pressure-test the narrative premium is to separate three questions that bubbles love to blend:

1. *Could* this be a huge market?
2. *Will* this company win a large share of it?

3. *At today's price,* are you already paying as if it has won?

The first can be true without the second. The second can be plausible without the third being a bargain. Bubbles thrive in the confusion between "big future" and "cheap today."

A small table that saves a lot of pain: translating story into numbers

Stories aren't the enemy. Unexamined stories are.

One way to keep your feet on the floor is to force a narrative to declare its terms. If the story is "this company will dominate," dominance must show up somewhere measurable.

Narrative claim	*What it must mean in numbers*	*Common failure mode*
"Total addressable market is massive"	Revenue growth that is both rapid and sustained; evidence the market is real, not just imaginable	TAM becomes a daydream; growth is assumed rather than demonstrated
"This is a winner-take-most business"	High and durable margins; strong retention; clear barriers to entry	Competition arrives; margins revert toward the industry mean
"Profits can come later"	A credible path to profitability; unit economics that work when growth spend is reduced	"Later" quietly moves further away; scale reveals hidden costs
"Network effects protect us"	Usage and value that grow together; switching costs that are real	Network effects are weaker than assumed; users multi-home
"The platform is inevitable"	Distribution advantage and pricing power without regulatory or customer backlash	Regulation, platform risk, or reputation risk caps pricing power

The point isn't to kill excitement. It's to make excitement pay rent.

Why fundamentals feel weak at the peak

If fundamentals are gravity, why don't they pull prices down immediately?

Because markets can stay social longer than you can stay patient.

At the peak of a bubble, fundamentals often look like they're improving, which makes skepticism emotionally expensive. Companies issue stock at high prices and raise piles of cheap capital. They hire aggressively. They buy competitors. They plaster their brand everywhere. The story seems to be "working," not just in price, but in reality. In those moments, the line between cause and effect blurs: are the fundamentals improving because the business is thriving, or because the market is subsidizing the business with an inflated stock price?

Even the language of "intrinsic value" can feel quaint when price is sprinting. A rising chart has a way of turning everyone into a philosopher. "Value is what people will pay." "The market is never wrong." "Old models can't see the future." These are not arguments; they are coping mechanisms for the discomfort of sitting out something that is making other people look smart.

But gravity doesn't vanish. It just takes longer to be felt when a crowd is leaning the other way.

A quick diagnostic: are you valuing a business, or joining a referendum?

There's a clean dividing line between investing and speculation, and it isn't morality. It's whether your expected return depends mostly on the business delivering cash flows, or mostly on someone else paying you more for the same asset.

Both activities exist in markets. The danger comes from pretending you're doing one while actually doing the other.

Ask yourself, with unpleasant honesty:

- If the stock market closed for five years, would I be thrilled to own this asset anyway?

- Can I explain, in one paragraph, the cash-flow engine of this business—who pays, why they pay, and what keeps them paying?
- If my forecast is wrong by 20% on growth or margins, does the valuation collapse?
- Would I still buy it if it dropped 40% tomorrow—not because it's "on sale," but because the underlying value is intact?

A bubble doesn't require you to answer "yes" to any of those. It only requires you to stop asking.

Actionable takeaway: a three-step anchor before you buy anything

Before committing money, force a quick anchor-and-story discipline. It takes 15 minutes and saves months of regret.

1. *Write the story in two sentences.* Not a manifesto. Two sentences. If it can't be said simply, it's probably not clear.
2. *Attach three numbers to the story.* Pick (a) revenue growth over the next 3–5 years, (b) a plausible long-run operating margin, and (c) a discount rate you can defend without using the phrase "it's different now." If you can't defend the numbers, you're not valuing—you're hoping.
3. *Name the narrative premium you're paying.* Compare today's price to a conservative valuation (even a rough one). If the price is far above it, don't argue with the gap—*label it*. Saying "I'm paying a 50% narrative premium" turns a hidden emotion into an explicit bet. Explicit bets can be sized. Hidden bets size you.

When the crowd is loud, anchoring yourself to cash flows isn't about being pessimistic. It's about refusing to confuse a great story with a fair price.

2.2 The Behavioral Trap: Biases, Herding, and FOMO

Picture the scene: you're not greedy, you're not reckless, and you're definitely not the type to gamble on "hot tips". Then you open your phone and see a friend's screenshot: a brokerage account up 62% in a month. Not a fund manager. Not a quant. Your friend from high school who once microwaved aluminum foil. The post has a chorus of comments—*What are you buying? Teach me. I'm in.*

In that moment the market stops being a market and becomes a social scoreboard. Your brain doesn't whisper, "Let's carefully evaluate expected cash flows." It asks a different question entirely: *Am I the only person not invited to this?*

That question is older than finance. It's older than money. It's the fear of being the lone fool while everyone else gets to be the genius.

Why smart people do dumb things in crowds

Bubbles don't require mass illiteracy. They require a particular kind of pressure: the pressure of watching other people get rewarded.

The human mind didn't evolve to price risk in the abstract. It evolved to read faces, track status, notice who is eating well, and avoid being cast out of the group. In markets, that wiring gets repurposed in a way that is both impressive and dangerous. We treat price movement like social information. We treat popularity like evidence. We treat a crowd as a substitute for certainty.

There's a common insult embedded in the way people talk about manias: "How could anyone believe that?" But belief is only part of the story. Plenty of people don't *believe* the valuations at the peak. They participate anyway, because disbelief doesn't pay their mortgage if the bubble keeps inflating for another year.

That's the behavioral trap in a sentence: emotion doesn't merely distort judgment; it changes the incentives you feel in your bones.

Social proof: the crowd as a fake credential

Social proof is a simple shortcut: if many people are doing something, it must be reasonable. In everyday life, it works often enough to be adaptive. If the restaurant is packed and the one next door is empty, you can guess which kitchen is better without reading a single review.

Markets weaponize this instinct. A rising price is a public, numeric signal of approval. It's not subtle. It updates every second. It's broadcast in green, which is not an accident; green is the color of "safe" and "go." When you watch an asset climb day after day, your brain starts treating the movement itself as proof that someone—someone smarter, richer, better connected—knows something you don't.

The most unsettling part is that this can be rational at the individual level. If you genuinely believe other traders are informed, following them may feel prudent. The problem is what happens when everyone follows everyone else. The market becomes a hall of mirrors where confidence is reflected, amplified, and mistaken for insight.

A classic bubble soundtrack includes the sentence: "I'm not saying it's worth this, I'm saying it's going higher." That isn't analysis. It's social proof turned into a strategy.

Information cascades: when your own eyes stop mattering

There's a specific crowd phenomenon that shows up again and again in manias: people stop trusting their private information once they observe enough public enthusiasm.

Imagine a line of investors walking into a theater. The first few have genuine information: maybe they understand the product, or they've read the filings, or they've built a model. The next group sees their buying and infers, "They must know something." The group after that sees a bigger crowd and infers, "All of them know something." Eventually, the decision to buy is not based on the underlying asset at all; it's based on the fact that buying has become the default behavior.

That's an information cascade: a process where observed actions overpower private judgment. The cascade is powerful because it feels like humility. You aren't claiming to be smarter than the group; you're admitting you might be wrong. You're deferring. In normal life, deference can be wisdom. In a mania, it's gasoline.

The fragility is the punchline. If the crowd's confidence is built on other people's confidence rather than on the asset's fundamentals, it can collapse from something surprisingly small: a missed earnings report, a policy change, a rumor that a big holder is selling. When the first person leaves the theater, the people behind them suddenly wonder what they missed. Exits can cascade the way entrances did.

FOMO: the emotion that turns patience into pain

Fear of Missing Out is often described like a silly modern disorder, as if it were caused by smartphones and cured by willpower. In markets, FOMO is more like a physical force. It can turn a sensible investor into an impulsive buyer, not because they forgot the risks, but because the social cost of doing nothing starts to feel intolerable.

FOMO has two sharp edges:

- It makes *time* feel like an enemy. If you don't act now, you'll lose your chance forever.
- It makes *prudence* feel like embarrassment. If you sit out and the asset doubles, you won't just miss money; you'll miss status.

That status element matters. A bubble is not just a price phenomenon; it's a conversation phenomenon. People swap tickers at parties. Coworkers compare returns at lunch. Family group chats turn into mini trading floors. Suddenly investing isn't private judgment; it's public identity. To abstain is to opt out of the shared excitement, like refusing a toast at a wedding.

FOMO can be so intense that it doesn't even require you to believe the story. It only requires you to believe that other people believe it long enough for you to profit.

Prospect theory: why the same person can be cautious and reckless

Standard finance likes neat preferences: an investor is "risk-averse," meaning they prefer a sure thing to a gamble with the same expected value. Real people are messier. They're cautious in some contexts and thrill-seeking in others, sometimes within the same week.

Prospect theory helps explain why. People tend to evaluate gains and losses relative to a reference point—a mental benchmark—rather than relative to total wealth. That reference point might be last month's account balance, the price you paid, your friend's returns, or the peak value your portfolio reached before it dipped.

Once you start keeping score that way, two predictable quirks appear:

- Losses hurt more than gains feel good. Losing $1,000 can sting like losing dignity; gaining $1,000 feels like a pleasant extra.
- People become risk-seeking when trying to avoid a loss relative to their reference point. After a drawdown, the temptation is to swing harder to "get back to even."

In a mania, the reference point often becomes the most toxic number imaginable: *the returns other people are posting*. If your friend is up 60% and you're up 8%, your brain doesn't register "8% is a fine year." It registers "I am losing." That sensation can push you into risk precisely when caution would be most valuable.

Then, when prices finally fall, the reference point shifts again. Investors anchor to the recent high and refuse to sell because selling would *lock in* the loss. They hold, hoping to be rescued by a rebound. What began as fear of missing out becomes fear of admitting you were wrong.

Overconfidence: the hidden engine behind ``I can time this"

Overconfidence is not the belief that you're the best investor alive. Overconfidence is the belief that you can control more than you can—that you can spot the top, dodge the crash, and exit gracefully, like a person stepping off a moving train without losing their hat.

The market punishes that belief with exquisite timing. It tends to reward overconfidence early, because in the inflation stage almost any exposure looks like skill. Gains arrive quickly and feel personal. You begin to attribute returns to your judgment rather than to the tide lifting all boats.

That's why bubbles are particularly dangerous for new investors. A few lucky trades can create a strong internal narrative: *I have a feel for this.* You trade more, you take bigger positions, you check prices more often, and you start confusing activity with control.

Empirically, frequent trading tends to be a wealth hazard for individuals once costs and bad timing are included. The tragedy is that the behavior feels like engagement and intelligence. You aren't passively watching; you're *doing something*. A bubble loves a participant. It doesn't care whether they're informed.

Herding: safety in numbers (until the numbers run for the door)

Herding is often portrayed as mindless imitation. In markets it can be more calculating than that.

If you manage money professionally—or even if you manage your family's money and feel responsible—being wrong alone is uniquely painful. When you lose with everyone else, it's a "market event." When you lose alone, it's a personal indictment. That asymmetry makes the crowd feel like cover.

Even outside professional investing, the same logic creeps in. Buying what everyone is buying offers a kind of emotional insurance. If it goes wrong, at least you weren't the only person fooled. If it goes right, you won't be the only person left out.

This is why bubbles can feel *safer* than normal markets while they're inflating. Price declines are quickly bought. Bad news is shrugged off. The crowd's presence creates liquidity—lots of willing buyers—and liquidity feels like safety. It isn't. Liquidity in a bubble is a party trick: it disappears when you need it most.

The darkest moment comes when the herd changes direction. Selling can cascade as quickly as buying did, and in leveraged markets it becomes forced selling: margin calls, redemptions, risk limits, lenders tightening terms. The psychology of "we're all in this together" flips into "get me out ahead of them."

Modern accelerants: group chats, screenshots, and manufactured urgency

Bubbles used to spread at the speed of newspapers, cocktail parties, and office gossip. Now they can spread at the speed of a forwarded message.

The mechanics are simple and effective:

- A persuasive personality posts a claim: an "inside source," a "whale wallet," an "algorithm," a "guaranteed pattern."
- The audience migrates to a more intimate channel: a private group chat, an invite-only server, an encrypted app.
- Trust is built with social proof: screenshots of wins, testimonials, carefully curated success stories.
- Urgency is injected: *this window is closing, don't miss the entry, we're about to run.*

Even when there is no fraud, the social design is the same: compress doubt, amplify excitement, and make waiting feel like stupidity. A screenshot of a gain is not analysis; it's an emotional weapon. It bypasses your skepticism and hits the primitive part of your brain that measures survival by relative standing.

The uncomfortable truth is that many people don't need to be tricked. They need to be given a narrative that feels communal and a clock that feels urgent. After that, they'll do the rest.

The greater-fool bargain: a contract nobody signs but everyone honors

The classic bubble logic is the greater-fool theory: it doesn't matter what something is worth as long as there's someone willing to pay more.

People often repeat this idea with a smirk, as if it were cynicism. In practice it's a fragile contract with the crowd. You are betting that enthusiasm will continue long enough for you to sell to the next buyer. But the crowd is not a single mind. It is a swarm of different time horizons, different levels of leverage, different degrees of conviction. Some participants will leave at the first wobble. Others can't leave because they borrowed to enter. The "greater fool" might not be greater. They might simply be later.

The terrifying part is that the strategy can work—for a while. That's why it's tempting. The market can validate the behavior repeatedly, making it feel less like speculation and more like a skill. A bubble doesn't punish you for being wrong early; it punishes you for being wrong *last*.

How to recognize FOMO in your own body before it hits your portfolio

FOMO isn't a spreadsheet error. It's a physiological state. Your pulse rises. Your attention narrows. You start looking for confirming opinions, not disconfirming ones. You check prices compulsively, as if vigilance itself could manufacture profit.

A useful diagnostic is to notice how your decision process changes:

- You stop asking, "What could go wrong?" and start asking, "How much could I make?"
- You feel irritated by cautious people, as if their caution were an insult.
- You look for reasons *today* is different from yesterday, because yesterday you didn't buy.
- You experience waiting as loss, not as discipline.

That last one is the signature. Waiting is normally neutral. In a mania, waiting feels like bleeding.

A practical defense: turn social pressure into a rule you can follow

Nobody permanently outgrows the crowd. The goal isn't to become a robot; it's to build a few guardrails that keep your worst impulses from getting the steering wheel.

Here is a concrete practice that works because it is slightly annoying:

1. *Write down what would change your mind.* Before buying, list two conditions that would make you sell or reduce: one fundamental (e.g., revenue growth stalls, margins deteriorate, a competitor undercuts pricing), and one market-based (e.g., the price falls 25% from your entry). If you can't articulate exit conditions, you're not investing—you're joining.
2. *Impose a waiting period on any "urgent" idea.* If an opportunity is framed as "must act now," enforce a 24-hour rule. Urgency is the favorite costume of bad trades. A quality long-term investment will still exist tomorrow; a crowd-induced impulse feels unbearable tomorrow, which is the point.

3. *Position-size as if you're wrong*. If the only way the trade works is by betting big, it's not conviction; it's desperation. Start smaller than your pride wants. A small position gives you psychological breathing room to think clearly.

The market will always offer you a chance to feel smart in public. The hard skill is learning to value being right in private more than being celebrated by the crowd.

2.3 Self-Fulfilling Prophecies: Reflexivity and Feedback Loops

A rising stock price does something uncanny: it can reach out of the screen and change the real world.

Not in a mystical way. In a *invoice-and-payroll* way.

A company with a soaring share price isn't just a company with happy shareholders. It becomes a company with cheaper money, better employees, friendlier headlines, and customers who assume it must be safe because "the market loves it." The chart becomes a credential. The valuation becomes a weapon. And for a while, the business may actually start to look as good as the price implied all along.

That's why bubbles are so persuasive from the inside. When you warn someone that prices are detached from fundamentals, they can point to improving fundamentals and say, "Detached? Look around." The loop is doing its job.

Reflexivity: when beliefs don't just reflect reality---they help create it

George Soros gave a name to a thing traders have always sensed: *reflexivity*. The idea is plain-language simple and emotionally uncomfortable: participants' beliefs can influence outcomes, and those outcomes feed back into beliefs. In finance, prices are not merely thermometers; sometimes they are thermostats.

In a physics experiment, you can observe a system without changing it. In markets, observing is acting. Buying pushes prices. Prices change financing conditions. Financing conditions change corporate decisions. Corporate decisions change business results. Business results justify the narrative that caused the buying. Round and round.

The temptation is to treat this as philosophical, but it's more like plumbing. The pipes are real, and the water pressure can blow the house apart.

The loop in everyday terms: the stock becomes a coupon

Start with a plain fact: companies can raise money in two main ways—borrowing (debt) and selling ownership (equity). Both become easier when your stock is expensive.

If your shares are priced at the market's idea of perfection, issuing new stock is like selling $100 bills you printed for $20 of economic sacrifice. Existing shareholders get diluted, yes, but the company receives a pile of cash at an unusually low cost. That cash funds marketing, hiring, acquisitions, research, and the kind of "strategic initiatives" that look brilliant when they work and inevitable when they're being pitched.

Debt becomes easier too, often for the same psychological reasons. Lenders, like everyone else, take comfort from prestige. A high market cap creates an aura of stability. Credit ratings can improve. Terms can loosen. In boom times, lenders don't merely ask, "Can you repay?" They also ask, "Will I look foolish if I say no while everyone else is saying yes?"

Now add a subtler mechanism: a surging stock price can become a recruiting tool. Equity compensation—stock grants and options—is a currency. When the currency is strong, you can buy more talent with fewer dollars of cash salary. Skilled employees who would never gamble at a casino will happily accept "upside" when it arrives wrapped in a job offer and a sense of belonging to the future.

So the price doesn't just summarize the company. It starts to *finance* the company.

The confidence channel: reputation as a form of capital

Not all capital is money. Some is trust.

When a company becomes a market darling, it earns a kind of reputational credit line. Customers take it more seriously. Partners return calls faster. Journalists write profiles that sound like destiny rather than probability. Regulators, at least at first, may be cautious about stepping in front of something that looks like national progress. Even competitors can hesitate, because competing against a hype machine is expensive and demoralizing.

This reputation can become self-fulfilling in small, practical ways. A consumer might choose a product because "they're the leader." A procurement committee might approve a vendor because "they'll be around." A CEO might sign a partnership because "that's where the world is going." None of these choices require fraud or irrationality. They only require the social shorthand that popularity equals safety.

It's a quiet transformation: the narrative premium becomes operating leverage.

Keynes's beauty contest: winning by guessing who others will admire

At the height of a mania, many investors aren't trying to estimate what a company is worth in the long run. They're trying to estimate what other investors will admire next week.

John Maynard Keynes compared the stock market to a beauty contest in which the prize goes not to the person who picks the face they find prettiest, but to the person who picks the face *everyone else will pick*. The contest becomes a game of second-guessing: what will the average opinion think the average opinion will think?

That sounds like a parlor game until you see how it changes behavior. It makes narratives more powerful than evidence. It makes timing feel more important than truth. It rewards the investor who can read the crowd's mood, not the investor who can read a balance sheet.

And it makes reflexivity turbocharged. If everyone believes that everyone else will pay more for "the leader in AI," then "the leader in AI" gets the cheapest capital, the best talent, and the loudest megaphone—which may actually help it become the leader. The market's belief doesn't just predict the outcome; it starts funding the outcome.

When high prices improve fundamentals (temporarily)

Here is the maddening part for anyone trying to stay disciplined: reflexivity means price can lead fundamentals, and the fundamentals can follow.

During the dot-com boom, many companies that were wildly overvalued still used their inflated stock to buy real assets: competitors, customer lists, software, data centers, brand recognition. Some acquisitions were smart, some were junk, but the mechanism was real. "Overvalued currency" is still currency.

During the housing bubble, rising home prices made households look creditworthy, which expanded lending, which further pushed demand, which further pushed prices. Homeowners extracted equity, spent more, and stimulated the economy in ways that made the boom feel justified. The narrative wasn't only told; it was lived.

In any bubble, the positive feedback loop can produce a genuine, measurable improvement in the short run. Revenues rise because demand is strong. Margins improve because the company can scale quickly. Losses can shrink because cheap capital subsidizes growth until profitability arrives. You can point to the income statement and feel vindicated.

It's not that skeptics were necessarily wrong about the endgame. It's that the market can create a *bridge period* where the bubble finances real progress, which makes the bubble harder to challenge. The loop buys time.

Time is what bubbles purchase with your skepticism.

The reflexivity diagram (and the trap hidden inside it)

Near the peak, it can feel as if fundamentals are catching up to price. Sometimes they are. But notice what the loop quietly assumes: that access to cheap capital and a glowing reputation remain available. The loop is powered by confidence. Confidence can evaporate faster than a company can cut costs.

The same mechanism that accelerates expansion also makes the eventual reversal sharper. When prices fall, equity becomes expensive. Recruiting becomes harder. Options go underwater. Debt covenants tighten. Customers get nervous. Partners hesitate. The company must shift from growth to survival at exactly the moment the market stops funding growth.

Reflexivity is not a free lunch; it's a leverage amplifier for sentiment.

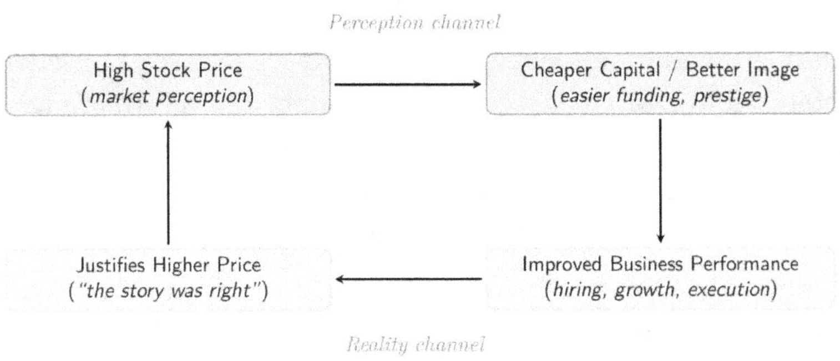

Why the loop is so hard to argue with (even when you're right)

Try debating a bubble participant who just got promoted at the company whose stock is soaring, or whose neighbor just sold a house for $200,000 above asking, or whose portfolio is green every day. They aren't responding to a theory. They are responding to a lived experience of validation.

Reflexivity makes that validation partially real. The capital raised is real. The hiring is real. The customer growth can be real. The sense of inevitability becomes real enough to influence behavior.

This is why bubbles tend to create a special kind of arrogance: not the arrogance of ignorance, but the arrogance of confirmation. You can be watching the loop and still feel like a fool for doubting it, because the loop keeps producing evidence that doubt is unnecessary.

A bubble doesn't just inflate prices. It inflates the social penalty for skepticism.

The moment reflexivity flips: from tailwind to trapdoor

A positive feedback loop is wonderful until it isn't. The flip often starts with something that seems small compared to the grand narrative:

- A funding round prices lower than expected.
- A central bank tightens policy and discount rates rise.
- A large holder sells, not because the story is false but because they need liquidity.
- A regulator asks an annoying question that forces the company to admit a constraint.
- Growth slows from "astonishing" to merely "good."

When the story is priced for perfection, "good" can be treated as failure.

Once prices fall meaningfully, the mechanism that once improved fundamentals can begin to erode them. Cheaper capital becomes expensive capital. "Brand halo" becomes "what do they know that we don't?" Employees refresh their resumes. Customers diversify vendors. Partners renegotiate. The company cuts spending, which slows growth, which confirms the market's new suspicion.

The loop doesn't stop. It changes sign.

This is the part that turns a bubble into a crash: when fundamentals don't merely fail to catch up to price, but start moving in the direction the falling price predicts.

A practical way to spot reflexivity at work: ask who is funding the dream

When you're evaluating a high-flying company, it's worth asking a slightly rude question: *Is the business improving because customers are paying, or because the market is paying?*

That doesn't mean market-funded growth is fake. It means it is conditional. It depends on continued confidence. A business with a genuine, self-sustaining engine can survive a funding winter. A business whose success is reflexively financed by its own valuation may discover, abruptly, that the valuation was the engine.

One of the cleanest indicators is what happens if you remove cheap money:

- If the company stopped raising equity tomorrow, could it still execute its plan?
- If debt became expensive, would the growth strategy still make sense?
- If stock-based compensation stopped feeling like a lottery ticket, would talent still stay?

Those questions feel harsh in boom times. They feel obvious later.

Actionable takeaway: write the ``without the stock price" version of the story

Before buying into a narrative that's already reflected in a surging price, do one exercise that cuts through the fog:

> *Write the bull case as if the stock price were flat for the next three years.*

No soaring valuation. No "momentum attracts capital." No assumption that the company can issue shares at will to fund growth. Just the business: customers, pricing, costs, competition, and execution.

If the story still works, you may have found a genuine compounding machine that the market happens to love.

If the story collapses without the stock price doing the heavy lifting, you've discovered reflexivity—and you now know what you're actually betting on: not just the company, but the crowd's willingness to keep feeding it.

2.4 Institutional Pressures: Agency Problems and Career Risk

When a bubble is roaring, the most misunderstood character in the story is the professional who *knows* it's a bubble and buys anyway.

From the outside, that looks like hypocrisy or incompetence. Inside the industry, it can look like a Wednesday.

A portfolio manager sits across from a risk committee, a CIO, or a client who is paying real fees for a simple promise: "don't embarrass me". Not "maximize long-run returns" in a philosophical sense. Don't be the only one who missed the rally. Don't be the person who underperformed the benchmark so badly that my board asks why I hired you. Don't make my peers laugh at a conference.

That's the first institutional truth that retail investors rarely see: markets aren't only driven by beliefs about value. They're driven by beliefs about *accountability*.

The agency problem: when ``good for the client'' isn't good for the manager

An agency problem is what happens when the person making decisions (the agent) doesn't bear the full consequences of those decisions, and doesn't receive the full rewards either. In investing, the agent is often the fund manager; the principal is the client.

The manager may genuinely care about the client—many do—but they also operate inside a career machine with its own incentives. Bonuses are often annual. Performance reviews are frequent. Clients can leave fast. Underperformance can trigger redemption requests, headlines, and awkward meetings. And the punishment is not symmetrical: one bad year can erase a reputation that took a decade to build.

So the manager's problem isn't simply "pick undervalued assets." It's "pick assets that will look sensible to other people on a quarterly timeline."

That's not cynicism. It's a constraint. And constraints shape portfolios.

Benchmark gravity: why the index is the real boss

Most professional managers are measured against a benchmark: the S&P 500, a growth index, a sector index, or a peer group. The benchmark is framed as a yardstick, but in practice it's a leash.

If the benchmark is 30% in tech and tech is melting faces, then a manager who holds 10% in tech is making a loud, controversial bet—even if they have a careful valuation case for why tech is overpriced. If they're right eventually but wrong this year, they don't get a trophy for moral courage. They get a client call that starts with, "Help me understand what you were thinking."

The index, in other words, is not neutral. It transmits the crowd's enthusiasm directly into the professional incentive system. As a sector rises, it becomes a larger part of the benchmark, which pressures managers to own more of it, which pushes prices further, which increases its benchmark weight again. The benchmark becomes a reflexive engine, not just a measurement tool.

If you ever wondered how "everyone" ended up owning the same expensive assets at the same time, start with the simple fact that their report cards were graded on the same curve.

Career risk: it's better to be wrong with the crowd than right alone

There's a line that survives because it's brutally accurate: it is safer to fail conventionally than to succeed unconventionally.

Career risk is the risk that you will lose your job, your bonus, your promotion track, or your client base *even if your investing logic is sound*—because you were early, different, and publicly wrong before you were privately right.

A manager who underperforms because they avoided the bubble can be fired *before* the bubble bursts. The clients don't experience that as "prudence." They experience it as "we paid fees and got left behind." Meanwhile, a manager who rides the bubble and loses money when it crashes can defend themselves with a shrug and a phrase that is strangely comforting to committees: "No one could have seen that coming." (They could have. But the committee can pretend otherwise.)

This is why sophisticated money can feed manias. Not because professionals can't do math. Because the payoff structure rewards survival through the next meeting.

The performance-chasing treadmill: why ``just a little exposure" becomes a lot

Institutional investing has a slow, steady way of turning a small compromise into a full position.

It starts innocently. A manager sees an overvalued sector and says, "I won't own much, but I can't be at zero." That becomes a token allocation. Then the sector keeps rising. The manager underperforms. Now the token allocation feels like a mistake. The next meeting looms. The manager increases exposure to "manage tracking error"—industry jargon for not looking too different from the benchmark.

At that point, the manager may still privately believe the valuation is absurd. But the portfolio has become a negotiation between belief and career. Every month the bubble continues, the negotiation moves in the bubble's favor.

This is how bubbles enlist professionals who would never describe themselves as trend-followers. They don't join the mania because they love the story. They join because the story has become part of their job description.

Herding with a suit on: reputations, peers, and the fear of looking stupid

Institutional herding doesn't always look like panic-buying. It often looks like a well-phrased slide deck.

A manager sits at a conference and hears the same names, the same themes, the same "must-own" companies repeated with slight variations. Analysts publish glowing notes. Competitors report strong performance. The business press frames the winners as inevitable. The manager feels a different kind of fear than the retail investor's FOMO. It's not "I'll miss a chance to get rich." It's "I'll look incompetent."

Reputational pressure is a powerful drug because it can be taken without feeling emotional. It feels like professionalism. If every serious investor owns the same handful of names, owning them feels prudent; not owning them feels like an act of stubbornness that will require explanation. Explanation is costly. Being unexplainable is fatal.

In a bubble, the crowd doesn't just provide social proof. It provides *cover*. Cover is valuable when you're managing other people's money.

Short horizons: the market's most expensive habit

A long-term investor can be patient with fundamentals. A short-horizon investor must be patient with the crowd.

Institutions talk about long-term investing the way people talk about going to the gym: with sincere aspiration and inconsistent attendance. In practice, many institutions have short horizons forced on them by the structure of their own business. End investors can redeem. Consultants can recommend changes. Boards can panic. The manager's time horizon is not "when value is realized." It's "until the next evaluation."

Short horizons create a perverse skill: knowing what other investors will care about *soon*. This shifts attention away from durable cash flows and toward narratives that can move the marginal buyer. It can also push managers toward liquid, crowded trades—the very trades that become most dangerous when everyone tries to exit at once.

A bubble loves short horizons because they turn investing into a continuous popularity contest. The manager isn't paid to be right in five years. They're paid to avoid being painfully wrong in the next two quarters.

The hidden client: the manager is also managing their own investors

There's another layer to the agency problem that almost never gets said out loud: a fund manager isn't only managing a portfolio. They're managing a *client's emotions*.

Clients claim they want long-term discipline. Many do. But in a mania, even good clients start to bargain: "We understand your valuation concerns, but could we just have *some* exposure? We're getting questions." The client isn't asking for returns; they're asking for reassurance.

The manager knows what happens when reassurance is denied. Assets leave. The business shrinks. The firm's leadership notices. The manager's autonomy shrinks. In that context, "riding the bubble" can feel like protecting the mandate, which feels like protecting the client—or at least protecting the ability to keep managing money at all.

This is how institutional logic turns a bubble into a group project.

Why ``smart money" sometimes makes the bubble worse

It's comforting to believe there's a class of investors who will step in, declare the price insane, and short the bubble back to reality. Sometimes that happens. Often it doesn't happen soon enough to matter.

Shorting is expensive and dangerous. Underperforming is career-threatening. Standing apart from the crowd requires not only conviction but endurance. In a prolonged mania, endurance becomes the rarest asset.

Meanwhile, the professionals who do participate bring something powerful: scale. Their buying is large, steady, and often rules-based. Their flows can dominate. Their research can become marketing. Their presence can legitimize the narrative, because if the big institutions are in, it must be real. Social proof with a logo is stronger than social proof with a screenshot.

This doesn't mean institutions are villains. It means incentives are. A bubble is a social and financial phenomenon, and institutions are social systems with paychecks attached.

A story you can recognize: the manager who ``can't be left behind"

If you want to see career risk in action, listen for certain phrases in professional commentary. They sound responsible. They can also be warning sirens:

- "We trimmed, but we can't be underweight."
- "Valuation is rich, but momentum is strong."
- "We respect the trend."
- "We're participating, but with risk controls."
- "Clients want exposure."

None of these statements are irrational. That's what makes them dangerous. They are the language of compromise, and compromise is how bubbles become crowded.

A manager who is truly insulated from career risk can say, "It's overpriced and we're out." Most managers are not insulated. They are employed. Employment is a constraint like gravity, just operating on a different dimension.

How reflexivity and institutions lock arms

Earlier we saw how rising prices can improve fundamentals by making capital cheap and prestige abundant. Institutions add another reinforcement: they create persistent demand.

When a stock enters major indices, passive funds must buy it. When it becomes a "must own" in model portfolios, flows follow. When a sector becomes a bigger part of benchmarks, active managers tilt toward it to avoid tracking error. When the managers tilt, prices rise, reinforcing the narrative that the leaders deserve their dominance. That narrative brings more capital, which improves the company's position, which validates the price. The loop is no longer just psychological. It is mechanical.

This is one reason bubbles can last longer than skeptics think they "should." The buying isn't always exuberant. Sometimes it's institutional housekeeping.

Actionable takeaway: when professionals sound calm, check the incentives

You can't see institutional incentives directly, but you can infer them. Before you take comfort in "the pros are buying," run a simple filter:

1. *Ask who is being measured against what.* Is the manager judged against a benchmark that is increasingly dominated by the bubble assets? If yes, their buying may be self-defense, not conviction.
2. *Watch for language of obligation.* Words like "can't," "have to," "must," and "need exposure" are tells. They signal constraint.
3. *Separate reputation from returns.* If an institution is praised for "participating" rather than for explaining cash flows, you're watching a career-risk market, not a value market.

The most useful skepticism isn't "professionals are stupid." It's "professionals are human, and humans respond to incentives." If you can identify the incentives, you can often predict the behavior—and in bubble markets, predicting behavior is half the battle of protecting yourself from it.

Chapter 3

The Anatomy of a Mania: From Boom to Bust

Why do financial markets consistently go from 'Goldilocks' perfection to 'Apocalypse' panic? We often treat crashes as unpredictable natural disasters, but history shows they follow a strict script. This chapter pulls back the curtain on the hidden mechanics of a meltdown, showing why smart people do reckless things when times are good, and how a single domino can knock down the global economy.

3.1 Stability Breeding Instability: The Minsky Cycle

A strange thing happens when nothing bad happens for a long time: people stop believing in bad things. Not intellectually-nobody stands up at a dinner party and announces, "I now reject the concept of recessions." It's subtler. It shows up in small choices that feel reasonable in isolation: the bank that loosens a covenant because the last five years of defaults were low; the CFO who refinances short-term because rates are cheaper; the homebuyer who figures the promotion is basically guaranteed; the investor who sells insurance against a storm that hasn't come in ages. Each decision is a pebble. Enough pebbles, and you get an avalanche.

Minsky's big idea—so simple it feels like it must be wrong the first time you hear it—is that stability is not the opposite of instability. Stability is often the *source* of instability. The calm period doesn't merely precede the accident; it helps *cause* it, because calm changes how humans price risk, how institutions compete, and how balance sheets are built. A financial system doesn't usually get fragile because everyone suddenly becomes reckless. It gets fragile because cautious people, responding rationally to a placid world, gradually redesign the system so it can't handle a rough one.

Think about driving on a long, straight, dry highway. The first few minutes, you're alert. You check mirrors. You keep a safe distance. But after an hour with no surprises-no sudden braking, no slick patches, no animals-your body relaxes. Your speed drifts upward. Your following distance shrinks. You reach for your phone at the stoplight because you've "earned" the right to multitask. The road didn't change; *you* did. Minsky's claim is that markets do the same thing: a quiet stretch trains participants to behave in ways that only work *during* quiet stretches.

The trick is that the danger doesn't announce itself as danger. It often announces itself as sophistication. It arrives dressed as "efficiency," "innovation," "risk management," "optimization." New products make old constraints feel old-fashioned. Competitive pressure makes restraint look like laziness. If one lender insists on a 30% down payment while the lender across the street offers 10%, the careful lender doesn't just lose business-they lose relevance. They begin to sound like someone warning about fire at a barbecue. Eventually they either loosen up or get replaced by someone who will.

Minsky's framework becomes clearest when you stop talking about "bubbles" as a mood and start talking about financing as a structure. Strip away the headlines and the memes and you can sort most borrowing into three buckets-three ways of answering a blunt question: *Can you pay what you owe without needing prices to keep rising?*

Hedge finance is the boring kind, which is exactly why it's healthy. Cash flows from the asset (or the borrower's income) cover both interest and principal on schedule. A homeowner with a fixed-rate mortgage they can comfortably afford from salary is in this camp. A business that can service debt from operating profits is, too. Hedge finance can still fail-job loss happens, profits slump-but it doesn't *require* a rising market to function. It's built to survive ordinary weather.

Speculative finance is a step closer to the cliff. Here, cash flows can cover the interest, but not the principal. The borrower is counting on being able to refinance, roll the debt, or sell the asset later to repay what's owed. Plenty of normal-sounding arrangements fit this description: a property investor expecting to refinance after renovations; a company funding long projects with short-term paper; a government rolling over debt as a permanent condition. Speculative finance can be stable as long as lenders remain willing to renew and markets remain liquid. That "as long as" is the thin ice.

Ponzi finance is the final transformation-and it's not named for criminal intent so much as for mechanical dependence. Cash flows don't cover interest *or* principal. The only way the borrower can meet obligations is by borrowing more, selling the asset at a higher price, or relying on the asset price to rise enough to justify ever-cheaper refinancing. This is where a boom starts eating its own tail: rising prices make refinancing easier, refinancing supports higher prices, and the whole machine begins to look like a perpetual-motion device-until it doesn't.

The unsettling point is that the system can slide from hedge to speculative to Ponzi without anyone feeling like they changed their personality. In fact, the shift often feels like prudence. When defaults are low, lenders loosen terms because "the data" says risk has fallen. When volatility is low, investors increase leverage because "the models" say the portfolio is safer. When prices rise steadily, borrowers rely on appreciation because "the market" has proven itself. Stability becomes evidence, and evidence becomes permission.

And then comes the moment that Minsky became famous for: the *Minsky Moment*. Not a date on a calendar, not a single headline, but the instant the crowd collectively discovers that yesterday's balance sheets only worked in yesterday's world. A small wobble-rates up a little, rents down a little, a few defaults in an ignored corner-exposes that many borrowers were never really paying their debts; they were refinancing them. The system, built for smooth roads, meets a pothole.

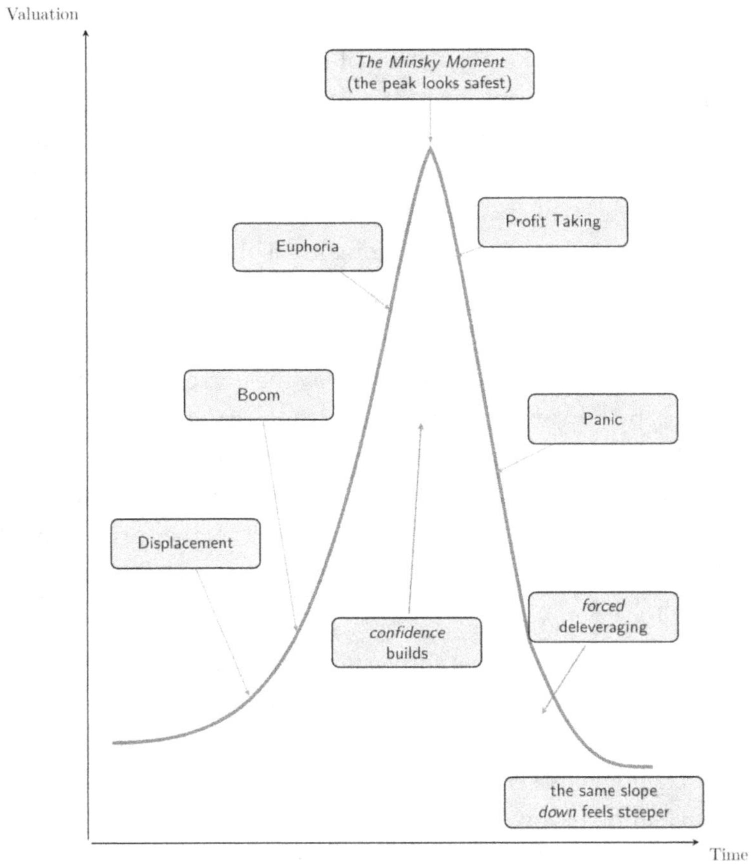

The rollercoaster image matters because it captures the emotional asymmetry. On the way up, nothing feels urgent. Gains arrive in increments. The daily habit becomes: *mark the position higher, feel smarter, repeat*. On the way down, time compresses. Decisions that were optional become mandatory. The market stops being a place where you choose what to do and becomes a place that tells you what you must do.

One reason Minsky's story rings true across eras is that it doesn't depend on any one technology or asset class. The props change-railroads in the 1840s, Florida land in the 1920s, dot-coms in the 1990s, housing in the 2000s, crypto in the 2010s and 2020s-but the plot is stubborn. A new opportunity arrives with a plausible narrative: productivity, democratization, a "permanent" shift. Early gains attract capital. Credit expands. The crowd learns the wrong lesson from good outcomes: "It worked" becomes "It's safe."

Here's a concrete way to see the shift from hedge to Ponzi without needing an economics degree: listen for the sentence that pays the bill. In hedge finance, the bill is paid by income: *my paycheck covers the mortgage*. In speculative finance, the bill is paid by access: *I can refinance when the teaser ends*. In Ponzi finance, the bill is paid by the market's approval: *I can sell to someone else at a higher price*. The sentence that pays the bill quietly migrates from the real world to the financial world-and once the bill is paid by the financial world, the system becomes allergic to disappointment.

That allergy creates a dangerous kind of optimism: optimism that must be constantly renewed. When a boom is powered by borrowers who need prices to rise, a flat market isn't merely boring-it's threatening. When a boom is powered by lenders who need refinancing to remain easy, small policy changes become existential. In that environment, the market starts to behave like an anxious animal: it looks calm until a noise triggers it, then it bolts.

The "displacement" phase is often misunderstood as pure hype, but it usually begins with something real. A genuine innovation appears: electricity, the internet, securitization, smartphones, a new trading venue, a new source of global savings. Or a new political regime changes rules. Or interest rates fall for long enough that old assumptions about financing costs start to feel silly. Displacement isn't a lie;

it's a new ingredient. The problem is what happens when the crowd decides the new ingredient makes the recipe foolproof.

Boom is where the financing structure starts changing under the surface. This is when bankers become "creative," which in practice means they become competitive in ways that erode the margin of safety. Covenants weaken. Underwriting standards "modernize." Maturities shorten because rolling debt seems effortless. Borrowers learn which questions lenders no longer ask, and lenders learn that the easiest way to grow is to stop asking.

Euphoria is where the story stops being about the asset and becomes about the self. People don't merely believe prices will rise; they believe rising prices prove something about their insight. That's when social proof becomes a financial factor. Your neighbor's gains are no longer background noise; they are a kind of accusation. Sitting out begins to feel like stupidity, not prudence. The crowd doesn't have to be irrational in a cartoonish way. It just has to be human: sensitive to regret, eager to belong, and intensely impressed by recent evidence.

The peril of euphoria is that it pushes the system into Ponzi finance without calling it that. A borrower who could have bought a smaller home with a safe payment structure instead buys a larger one because "equity will build fast." A fund that could have accepted modest returns instead borrows to "enhance" them because volatility is low and lenders are eager. An institution that could have held liquidity instead holds less because cash is "idle." From inside the boom, these look like smart upgrades. From outside, they look like the gradual removal of seatbelts.

Profit taking is the phase that sounds healthiest but often arrives too late to save the system. Early sellers are mocked until they're envied. And when enough people try to be the adult in the room, they discover the room is built on leverage. Selling doesn't just change ownership; it changes collateral values. It changes funding. It changes whether refinancing is possible. In a highly levered environment, "taking some chips off the table" is not a neutral act. It tugs on the structure.

Panic is not merely fear—it's geometry. It's what happens when everyone discovers that the exits are smaller than the crowd. When financing is fragile, a price drop is not just a loss; it is a threat to solvency, a trigger for forced action. That's why crashes feel like betrayal. People think they're panicking because they're scared; often they're panicking because they're constrained. They sell because they must, and then they tell themselves a story about why they chose to.

Minsky's brilliance was to insist that the "must" is endogenous. It's built during the good times. When volatility is low, leverage feels safe, which encourages more leverage, which makes the system sensitive to any return of volatility. When defaults are low, lending terms loosen, which increases borrowing, which increases exposure to any return of defaults. Calm doesn't just precede the storm; it trains the architecture to fail in it.

The crowd is often blindsided because the most dangerous part of the cycle is invisible in a rising market: the balance sheet. Prices going up act like a flattering mirror. They make borrowers look solvent, lenders look skilled, and risk models look prophetic. That's why warnings are so hard to sell. If someone points out that a system is moving toward speculative or Ponzi finance, the comeback is always the same: "But look—everything is fine." The boom becomes its own proof.

The practical value of Minsky's lens is not that it predicts the date of the next crash. It doesn't. The value is that it tells you what to watch when everything looks rosy. Not the headlines. Not the price chart. The financing terms. The reasons people give for why debt is "different now." The casualness with which refinancing is assumed. The way risk gets renamed as "efficiency." If you want a single phrase that captures a Minsky-style warning sign, it's this: *When success becomes the justification for fragility, the system is already changing shape.*

A useful way to apply this personally is to treat your own finances like a small bank-because, functionally, you are one. Your income is funding. Your savings are capital. Your debt payments are obligations. Your ability to refinance is not a right; it's a market condition. During calm periods, ask an unfashionable question: *If the refinancing window closed for two years, would my plan still work?* If the answer is no, you're relying on stability to remain stable, which is exactly the bet Minsky warned about.

Actionable takeaway: When markets (or your personal finances) feel safest, run a "boring stress test" focused on cash flow, not optimism. Write down every major obligation you have over the next 12 months-mortgage/rent, car payments, credit cards, student loans, business expenses-and circle the ones that you can only meet by rolling debt, selling assets, or assuming prices keep rising. Then choose one concrete de-risking move you can execute this month: refinance into fixed payments, extend maturities, build a cash buffer, or reduce leverage. The goal isn't to be fear-driven; it's to stay in hedge finance even when the crowd insists it's outdated.

3.2 The Fuel of the Boom: Leverage and Margin

If you want to know whether a market rally is merely cheerful or quietly dangerous, don't start with the price chart. Start with a more intimate document: the loan agreement. Manias are not powered by optimism alone. Optimism is the spark; borrowed money is the oxygen. With enough oxygen, even a small flame becomes a house fire—bright, fast, and hard to control.

Leverage is the plain-English word for an old temptation: using other people's money to magnify your own results. It works because returns are measured on *your* equity-the cash you actually put in-while the asset you're buying is bigger than your equity. When the asset rises, you keep most of the upside. When it falls, you absorb most of the damage. Leverage is a financial amplifier that doesn't care whether the signal is music or noise.

A quick thought experiment shows why the boom always feels so persuasive. Suppose you buy a house for $500,000. If you pay cash, a 10% rise in prices makes you $50,000 richer. Nice. Now suppose you buy the same house with 10% down: $50,000 of your money and $450,000 borrowed. If the home rises 10%, you still gain $50,000 in home value-but now that $50,000 gain is on a $50,000 equity base. You just doubled your money (before costs). The house didn't become a better house. You just put it on stilts.

Those stilts are what make leverage so addictive. The early stage of a boom rewards aggressiveness with what looks like genius. People don't merely profit—they profit *fast*. Fast profits are socially radioactive: they spread. They make cautious investors look like they're failing a test everyone else passed. They also do something more dangerous than envy. They create a new baseline of "normal" returns, and once your brain rewires itself to expect 20% a year, 7% starts to feel like an insult.

Now take the same house and run the tape backward. Prices fall 10%. The cash buyer loses $50,000—painful, but survivable. The 10% down buyer loses $50,000 too, because the asset fell by the same amount. But the down payment was only $50,000. That's 100% of the equity erased. The borrower hasn't merely had a bad year; they've been knocked out of the ring. Leverage turns small declines into existential events.

That's the first lesson: leverage changes the *shape* of risk. It's not a smooth dial from "safe" to "risky." It's a trapdoor. You can enjoy months of calm and then, with one ordinary price move, find yourself in a different universe-one where you no longer have the option to wait for a recovery.

The second lesson is that leverage doesn't just magnify outcomes; it rewrites behavior. People with low leverage can afford to be patient. People with high leverage become *time-sensitive*. They must care about the next week, the next margin check, the next refinance window. This is why highly levered booms are so prone to sudden mood swings. The crowd isn't fickle by nature; it's being pushed around by financing constraints that don't negotiate.

Margin is the stock-market version of that house example. A *margin account* lets you buy securities with borrowed money from your broker. Your broker fronts part of the purchase price, and your securities serve as collateral. Your *equity* in the account is simply what you own minus what you owe: the market value of your holdings minus the margin loan. That definition is almost insultingly simple-which is exactly why people get into trouble. They focus on the market value ("My portfolio is $200,000!") and forget the subtraction ("...and I borrowed $120,000 to get there.").

Margin feels civilized during a rising market because it produces a clean, flattering arithmetic. Suppose you have $50,000 and you buy $100,000 of stock on margin. A 20% gain turns $100,000 into $120,000. After repaying the $50,000 you borrowed (ignoring interest for simplicity), your equity becomes $70,000. You didn't just gain 20%; you gained 40%. That extra 20 percentage points feels like free skill.

But leverage is not free skill; it's rented horsepower. The rent is paid in fragility.

The fragility shows up in two rules that sit quietly in the background until they suddenly run your life: *initial margin* and *maintenance margin*. Initial margin is how much of the purchase price you must put up at the start. Maintenance margin is the minimum equity you must keep in the account as prices move. You can think of initial margin as the down payment and maintenance margin as the "keep the collateral healthy" requirement.

Here's the crucial detail: maintenance margin is not a warning label. It's a trigger. If your equity falls below the requirement, you can get a *margin call*—a demand to add cash or securities, or for the broker to reduce your position. And the broker is not a patient therapist. Brokers typically reserve the right to protect themselves quickly. In practice, that can mean forced selling at precisely the worst time, when the screen is red and liquidity is thin.

A margin call is where the boom stops being a story about opinions and becomes a story about plumbing. You may believe your investment is undervalued. You may believe the market is overreacting. You may even be right. None of that matters if you can't meet the call. In a levered system, the question "What is this worth?" can be temporarily replaced by "Who can hold it?" That's why leverage turns markets into endurance contests, and why endurance is often determined by financing terms, not intelligence.

It gets worse: maintenance requirements are often not fixed in spirit even when they are fixed on paper. Brokers can impose "house" requirements—tighter rules than the minimum—especially when volatility rises or a stock becomes hard to borrow or hard to trade. The same asset can become more expensive to finance overnight, not because its cash flows changed, but because the lender's risk tolerance did. This is the financial equivalent of your landlord demanding three months' rent upfront—right after you got laid off.

The most brutal feature of margin is that it can force you to sell low without ever giving you the "choice" to be wrong later. If you're unlevered and the price falls, you can wait. If you're levered and the price falls, you may be sold out. That difference—*optional* versus *mandatory*—is where booms are won and crashes are accelerated.

Leverage also doesn't stay politely contained within obvious borrowers. It leaks into the system through institutions that don't look like banks but behave like banks. This is where the phrase *shadow banking* earns its eerie name. Shadow banking is the network of lending and borrowing that happens outside traditional deposit-taking banks: money market funds, repos (short-term secured loans), hedge fund prime brokerage, securitization chains, finance companies, and a zoo of vehicles created to hold assets while funding them with short-term liabilities. The shadow system often grows fastest in good times because it can offer what regulated banks struggle to offer: speed, flexibility, and the illusion that risk can be sliced thin enough to disappear.

The illusion is not that risk is gone. The illusion is that it's been domesticated.

A traditional bank is constrained by capital requirements, supervision, and the nuisance of depositors who can panic. Shadow banking is constrained by something less formal and more twitchy: confidence in collateral. Much of shadow funding is collateralized-meaning loans are backed by securities that can be seized and sold if the borrower fails. In calm markets, that sounds safer than unsecured lending. In stressed markets, the "safety" becomes a guillotine. Lenders demand bigger cushions (*haircuts*) against collateral. Borrowers must post more collateral or shrink positions. Shrinking positions means selling assets. Selling assets pushes prices down. Prices down make collateral look worse. The haircut rises again. The guillotine falls in slow motion, then all at once.

This is why leverage is not merely a personal risk choice; it's a system property. If one investor borrows, that investor is exposed. If *everyone* borrows, the market's behavior changes. Price declines no longer produce bargain hunters alone; they also produce forced sellers. The presence of forced sellers makes downturns steeper and faster, which makes lenders more nervous, which makes them tighten terms, which creates more forced selling. The accelerant doesn't merely make the fire hotter; it changes the direction the flames travel.

Leverage also hides in a place people rarely look: the *maturity* of the debt. A company can be "low leverage" in a headline sense-debt-to-equity ratios that look fine-and still be fragile if the debt must be rolled every few weeks. Short-term funding is cheap in good times for the same reason umbrellas are cheap in a drought: nobody feels the urgency. But short-term funding turns a long-term bet into a daily referendum. If lenders decide not to renew, the borrower must sell assets quickly. Quick sales create bad prices. Bad prices create insolvency. A balance sheet can look healthy right up to the hour it's forced to liquidate.

That's the nightmare scenario in a panic: not merely losses, but *losses synchronized with funding stress*. It's one thing for prices to fall. It's another for financing to evaporate at the same time. When both happen together, the market begins to behave like a building whose stairwells are narrowing as the smoke thickens.

During booms, leverage is often defended with sophisticated language: "risk parity," "carry," "relative value," "enhanced yield," "capital efficiency." Sometimes the strategies are genuinely thoughtful. The trap is that sophistication can create a false sense of control. Leverage turns many risks into the same risk: *liquidity risk*—the risk that you cannot get cash when you need it, at a price you can survive.

A vivid modern example is the pre-2008 housing machine, which was less a single bubble than a factory that converted mortgages into securities, then used those securities as collateral for more borrowing. Homebuyers used leverage through small down payments. Mortgage lenders used leverage through warehouse lines and securitization pipelines. Wall Street used leverage through repo funding and balance-sheet expansion. Ratings models and diversification stories acted like insulation around a live wire. When house prices stopped rising, the system didn't merely "reprice." It seized, because refinancing and rolling short-term funding were the invisible assumptions holding the structure up.

The reason this matters for any future mania—tech, housing, commodities, crypto, something not yet invented—is that the asset is rarely the true accelerant. The accelerant is the financing that allows the asset to be owned by people who can't actually afford to hold it through a drawdown. If you want a bubble that can pop violently, you need a market where many participants are effectively renting their positions from lenders, with the right to stay only as long as prices behave.

There's also a quieter way leverage distorts incentives: it turns "not losing" into "not being caught." A fund manager who uses leverage and gets lucky can show spectacular returns and attract more capital. The reward comes quickly. The penalty for hidden tail risk often comes later, sometimes after bonuses have been paid and careers have moved on. That mismatch—fast reward, delayed consequence—is a kind of moral leverage. It encourages risk-taking even among people who are not thrill-seekers, because the career math can be cold: underperform for three years and you're fired; outperform for one year and you're a star.

This career pressure is why leverage blooms in crowds. Individuals may understand the risks perfectly well and still choose to participate because the local incentives are brutal. In a boom, caution can feel like professional malpractice. The irony is that leverage is often marketed as a way to "manage" risk—hedging here, diversifying there—while it is simultaneously increasing the one risk you can't hedge easily: the risk that the whole system tries to exit at once.

So what does a leverage-fueled boom look like in real time, before it breaks? It usually looks like ease. Credit is available with less documentation. Promotions emphasize monthly payments rather than total cost. New products offer "flexibility," which often means the ability to postpone pain. Spreads compress. Volatility falls. People begin to treat low volatility as proof that the world has become less volatile, rather than proof that everyone is leaning the same way.

And when the turn comes, the first sign is often not a dramatic crash. It's a small tightening: lenders get picky, haircuts rise, a few margin calls appear, a refinancing deal fails, a "temporary" funding market hiccups. The public may not notice. The people living on borrowed oxygen notice immediately, because their positions start to feel heavy.

A practical way to think about leverage is to separate the *economic* bet from the *financing* bet. The economic bet is: "Is this asset likely to be worth more in five years?" The financing bet is: "Can I survive the path between now and five years?" Leverage often doesn't change the economic bet. It changes the financing bet from a footnote into the main event. People who blow up are frequently right about the long-term story and wrong about the short-term funding.

If this sounds abstract, translate it into a personal rule: leverage is a promise you make to your future self under uncertainty. In good times, you'll feel proud of the promise. In bad times, you'll resent it like a trap you laid for yourself.

Actionable takeaway: Before you borrow to invest-whether it's a margin loan, a HELOC used to buy stocks, a short-term loan funding a long-term asset, or a "low payment" deal that depends on refinancing-write down two numbers on a piece of paper: (1) the price decline that would wipe out your equity, and (2) the cash you could produce within 72 hours without selling the investment. If a normal-sized drawdown in that market gets you close to (1), or if your 72-hour cash is thin, you're not merely taking market risk-you're taking *forced-selling* risk. Reduce the leverage, lengthen the funding, or increase the cash buffer until you can survive being wrong for a while. In manias, survival is the edge that looks boring right up until it looks brilliant.

3.3 The Tipping Point and Panic Dynamics

The moment a bubble breaks rarely looks like a movie scene. There's usually no single villain, no one sentence that detonates the market, no clean bell ringing at the top. More often it begins as a minor inconvenience: a refinancing that takes longer than expected, a dealer who quotes a slightly wider spread, a lender who asks for "just a bit more collateral," a fund that gates redemptions for "operational reasons." In a healthy system those are speed bumps. In a leveraged system they're the first hairline cracks in a dam.

The cartoon version of a crash is emotional—panic, hysteria, irrational selling. The real version is frequently mechanical. People sell because they are forced to sell. They sell because the phone call arrives, or the automated risk system flashes red, or a lender simply declines to roll financing. The emotion comes after the mechanics begin, like thunder after lightning. If you want to understand why prices can fall so far, so fast, you have to understand how quickly "I want to hold" turns into "I can't hold."

Wile E. Coyote doesn't fall when he runs off the cliff. He falls when he looks down. Financial markets have the same delayed physics. In the late stage of a boom, participants can feel the air thinning—valuations stretched, stories strained, cash flows lagging—but the system keeps moving because funding is still available and marks are still friendly. Then something changes that makes everyone *look down*: volatility rises, a previously liquid market becomes sticky, a few defaults appear in a place that was supposed to be "contained." The realization hits all at once: the only reason the structure looked stable was because it was being continuously refinanced.

The tipping point is often described as a "loss of confidence," which makes it sound like a mood swing. But confidence in markets is not just a feeling; it is a set of permissions. When confidence is high, lenders permit leverage. Dealers permit tight bid-ask spreads. Investors permit themselves to believe today's price is a fair approximation of value. When confidence weakens, those permissions are revoked. The system becomes stricter at the exact moment it needs to be forgiving.

There are two kinds of liquidity that matter in a panic, and confusing them is one reason crashes seem like black magic. *Market liquidity* is the ability to sell an asset without moving the price too much. *Funding liquidity* is the ability to borrow against that asset—to get cash when you need it. In calm times they reinforce each other upward: easy funding brings in buyers, which makes markets liquid, which reassures lenders. In stress they reinforce each other downward: funding is pulled, forced selling rises, markets become illiquid, which scares lenders, which tightens funding again. The doom loop is not a metaphor. It is a feedback mechanism.

Margin calls are the most visible trigger, because they are blunt. Your broker (or your lender, or your prime broker, or the repo desk) isn't debating philosophy; they're managing collateral. When the value of the collateral falls, the lender demands more protection. That demand shows up as a margin call: add cash or reduce the position. In other words, pay now or sell now. In a fast decline, many investors receive the same ultimatum at roughly the same time. It's a synchronized shove.

Forced selling is the part people underestimate because they imagine selling as a voluntary act. In a panic, selling is often a medical reflex, not a decision. Institutions have risk limits. Funds have leverage caps. Banks have collateral agreements. Even a disciplined investor can become a forced seller if their financing requires it. The market doesn't ask whether you're "long-term." It asks whether you can post collateral on Thursday.

Fire sales take the mechanics and add gasoline. A *fire sale* is selling not because the price is attractive, but because you need cash *now*. That urgency changes price formation. Bids disappear, not because everyone believes the asset is worthless, but because everyone understands that the seller is distressed—and nobody wants to be the first bidder in a falling elevator. The first trades print at ugly levels, those levels become new marks, and suddenly more balance sheets look weak. The fire sale creates its own justification.

Here's the brutal arithmetic: the more levered the system, the less it can tolerate volatility. A one-day move that would be a shrug in a cash market becomes a crisis in a margined market. Volatility makes lenders nervous. Nervous lenders demand more margin or bigger haircuts. Bigger haircuts force deleveraging. Deleveraging creates more selling, which creates more volatility. The panic feeds on the very thing it produces.

The psychological shift is real, but it doesn't start as simple fear. It starts as a kind of insult: "Wait—this thing that always traded easily is suddenly hard to sell." That's when traders and executives stop thinking in terms of return and start thinking in terms of survival. The language changes. Conversations that were about upside ("how much can we make?") become about funding ("how much cash do we have?"). The room temperature changes. People stop laughing. A market can go from casual to feral in a week.

Crowd psychology then piles onto the mechanical cascade. Humans are exquisitely sensitive to social signals under stress. When prices fall, we look for an explanation. When we can't find a satisfying one, we assume someone else knows something we don't. That's how the selling becomes contagious even among investors who are not levered. Nobody wants to be the last person holding the bag, especially when the bag is leaking.

Contagion is one of the most misunderstood words in finance because it sounds mystical, like panic telepathy. The transmission mechanism is usually mundane: shared assets and shared funding needs. If many institutions own similar "safe" assets, and a subset of them is forced to sell, the price drops for everyone. Those price drops then trigger risk limits and margin requirements for others, even if their original positions were sensible. One firm's liquidity problem becomes another firm's solvency question through the simple act of marking to market.

The 2008 crisis offered a particularly sharp lesson in non-bank contagion: money market mutual funds. They look like parking lots for cash—conservative, boring, designed to feel like a checking account with a slightly better haircut. In mid-September 2008, after Lehman Brothers failed, the Reserve Primary Fund's net asset value fell below $1.00, driven by its exposure to Lehman debt. That break of the buck mattered less for the dollars involved than for the psychology: the market realized that "cash-like" could mean "not cash." The fund then faced roughly $40 billion of redemption requests in two days, and investors pulled about $300 billion from prime money market funds over the week of September 15, 2008. Those funds met redemptions by selling assets into stressed markets, which pressured prices and strained funding further. Panic didn't travel by rumor; it traveled by redemption requests and forced liquidation.

That episode is worth sitting with because it shows how modern panics can bypass the obvious icons like bank branches. The run wasn't a line of depositors outside a marble building. It was a digital stampede through "safe" funds, transmitted by clicks, and amplified by the need to sell the same high-quality short-term instruments that everyone else was also trying to sell. When the crowd wants liquidity, it often demands it from the same small set of assets. Those assets then stop being liquid, precisely because they're the ones being used as the escape route.

Panic also changes what information means. In normal markets, bad news is digested, debated, and priced. In panic markets, information is treated like a weapon. People stop asking "Is this true?" and start asking "Will this make someone pull funding?" Rumors that would be shrugged off in good times become decisive in bad times because the real fear isn't the rumor—it's the reaction.

This is why "cheap" assets can keep getting cheaper. Traditional value investors like to say, "Be greedy when others are fearful." That advice is emotionally brave but mechanically incomplete. In a liquidity spiral, fear is not just sentiment—it's a reflection of binding constraints. Prices can overshoot because the marginal seller is distressed and the marginal buyer is cautious. When the marginal buyer disappears, prices are set by whoever is desperate enough to hit the bid.

The market's microstructure—how trades actually happen—also becomes visible during panics in a way it never is during booms. Bid-ask spreads widen. Depth evaporates. Correlations go to one: assets that are supposed to diversify suddenly fall together because investors sell what they *can*, not what they *want*. It's common to see high-quality assets sold alongside low-quality ones because liquidity is being rationed. The selling is about raising cash, not expressing an opinion about fundamentals.

One of the cruelest dynamics is that the system can punish even prudence. Suppose you run a conservative portfolio, but you hold the same "safe" collateral as everyone else. If others dump it to raise cash, the price drops. Your marks fall. Your VaR or internal risk limits tighten. You're forced to reduce exposure, perhaps selling the same collateral at the same bad prices. You didn't do anything reckless. You just stood in a crowded room where someone yelled fire.

And then, because humans can't tolerate randomness, the narrative machine kicks in. People begin to retrofit explanations: "It was obvious," "the signs were everywhere," "anyone could have seen it." That post-hoc certainty is comforting but dangerous. It suggests that panics are primarily failures of foresight, when many are failures of structure. The key question is not "Who should have known?" but "Who was forced to act?" In a leveraged system, the forced actors set the tempo.

All of this helps explain why authorities often respond to panics by focusing on liquidity—backstops, guarantees, lending facilities—because when the doom loop is running, injecting funding liquidity can prevent the spiral from feeding on itself. That doesn't mean every intervention is wise or costless. It means the problem in the acute phase is often not valuation but financing. You can't reason a system out of a margin call. You have to meet it, delay it, or change the rules that generate it.

The uncomfortable takeaway is that panics are not rare meteor strikes. They are an emergent property of leveraged crowds. The late-stage boom builds a world where many actors must sell into weakness. Once enough of them are forced, the decline becomes self-reinforcing. The crowd's fear is not a glitch—it's the emotional reflection of a financial architecture that has stopped giving people time.

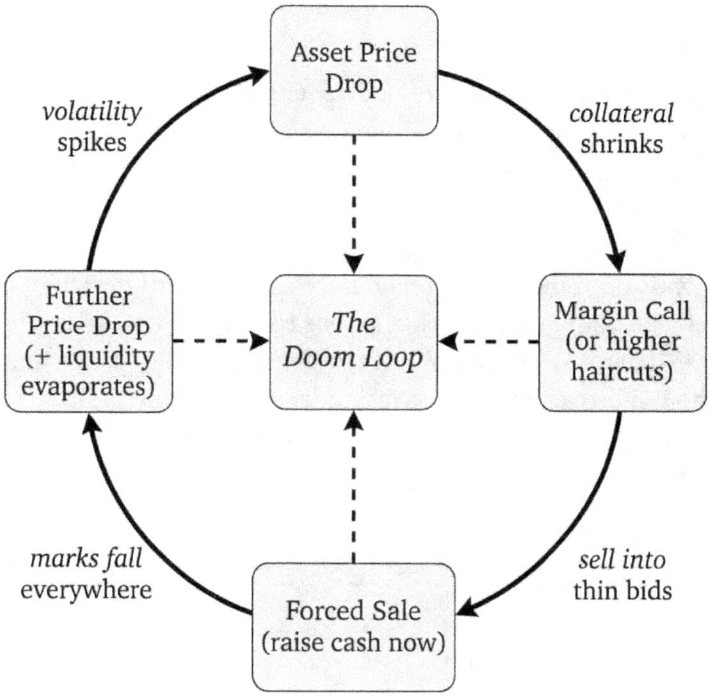

The diagram looks almost too neat, which is why it's useful. Real panics are messy—multiple assets, multiple lenders, multiple feedback loops—but the core cycle repeats with depressing consistency. Once you see the pattern, you start noticing the early symptoms: rising haircuts, widening spreads, "temporary" restrictions, liquidity hoarding, and the peculiar phrase that often appears right before trouble: "out of an abundance of caution."

For an individual investor, the big mistake is assuming you'll be able to behave like a calm, rational adult during a panic while using tools that remove your ability to wait. If you're financed overnight, if your portfolio can be liquidated by someone else, if your obligations require market access at specific times, your future behavior is being pre-committed by your financing. You're not just betting on the asset. You're betting on the market's willingness to fund the asset during chaos.

Actionable takeaway: Identify your personal "forced seller" triggers and eliminate at least one. Concretely: write down any position or obligation that could compel you to raise cash on short notice (margin debt, short-term borrowing, a HELOC used to invest, concentrated holdings that could gap down, or a business cash cycle that depends on rolling credit). For each, ask one hard question: *If this asset fell 25% in a week, would anyone else have the power to force my hand?* If the answer is yes, reduce the leverage, lengthen the funding, or hold more cash than feels fashionable. In a panic, cash isn't "dry powder." It's *time*—and time is what the doom loop steals first.

3.4 The Aftermath: Deleveraging and Restructuring

When the crash is fresh, the damage looks like a price chart. Six months later, it looks like a kitchen table. A couple staring at a stack of envelopes, deciding which bill gets paid first. A small business owner refreshing a bank portal, praying the credit line hasn't been cut. A newly cautious manager canceling a hiring plan not because demand vanished overnight, but because the company's balance sheet suddenly feels like thin ice.

The aftermath of a mania is not a single event. It's a long, grinding change in behavior—less borrowing, less spending, more saving, more scrutiny, more suspicion. The boom was a social mood. The bust becomes an accounting exercise.

Deleveraging is the clean word for an ugly process: paying down debt, restructuring debt, defaulting on debt, or inflating it away. In the boom, leverage felt like power. In the bust, leverage feels like gravity. Every decision is measured against a burden that does not politely shrink when your income or profits shrink. In good times, debt is a tool. In bad times, it becomes a schedule.

The first shock after a crash is obvious: wealth evaporates, confidence collapses, and the financial press starts using the word "bloodbath" with a straight face. The second shock is quieter and often more important: the balance sheet becomes the boss. A family that was willing to splurge when their home equity was rising becomes conservative when that equity disappears. A company that used to finance growth with easy credit starts hoarding cash. A bank that used to compete aggressively starts acting like a museum curator, touching nothing without gloves.

This is where the economy can get stuck in what's often called a *balance sheet recession*. It's an unintuitive idea at first because recessions are usually explained as drops in demand: people buy less, businesses sell less, unemployment rises. A balance sheet recession is demand weakness with a specific cause: lots of households and companies decide, simultaneously, that their top priority is repairing their finances rather than expanding their lives. They pay down debt instead of buying stuff. They refinance instead of hiring. They cut capex instead of building factories. The result is a paradox: what is prudent for each individual becomes painful for the whole economy.

It's the thrift trap in the real world. If one household saves more, it becomes safer. If *everyone* saves more at once, businesses lose revenue, layoffs rise, incomes fall, and the attempt to save becomes harder. The economy is not a morality tale where virtue is always rewarded quickly. Sometimes virtue is rewarded eventually, and punished immediately.

The aftermath also has a peculiar emotional texture. During the boom, risk felt like a game. After the bust, risk feels like shame. People don't merely lose money; they lose face. The most painful losses are often social: the investor who convinced friends to buy, the entrepreneur who expanded at the top, the family that stretched for a house and now can't move. This is why post-crash periods are often defined by caution that borders on trauma. The goal becomes not "make money" but "never feel that again."

In the early part of recovery, the market can bounce sharply—sometimes violently—because panic selling ends, liquidity returns, and bargain hunters appear. Those rallies can be convincing. They can create the sense that the crisis was a temporary fever. But a balance sheet problem is not cured by a better mood. A household with too much debt is not rescued by a 15% stock-market rally if their paycheck is smaller. A bank with questionable assets is not made healthy by optimism; it's made healthy by capital, write-downs, and time.

Debt overhang is the phrase economists use for the same heavy feeling: when debt is so large relative to income or profits that it discourages new spending and investment. A company with a debt overhang may have good projects available—new products, new factories, acquisitions—but hesitate because the benefits of success go first to creditors. When the future is being used to pay for the past, ambition becomes expensive. You can see it in corporate behavior: buybacks replaced by balance-sheet repair, growth targets replaced by "disciplined capital allocation," expansion replaced by "focus on core."

For households, the debt overhang shows up in more ordinary language: "We can't." Can't take the vacation. Can't fix the roof. Can't afford daycare without juggling. Can't move because the mortgage is underwater. Can't change jobs because health insurance is attached. A bust doesn't just reduce spending; it reduces mobility. It keeps people in the wrong homes, the wrong jobs, sometimes the wrong cities. The aftermath is as much about friction as it is about fear.

A crash also rearranges the pecking order of spending. When money is tight, spending shifts from optional to essential. Restaurants feel it. Travel companies feel it. Luxury brands feel it. But so do state and local governments when tax revenue falls. In a bust, you don't just delever. You reprioritize. And when millions of people reprioritize in the same direction, the whole economic landscape changes.

There is a darker dynamic that can lengthen the pain: debt-deflation. The idea is simple and vicious. When asset prices fall, collateral values fall. When collateral values fall, lenders tighten and borrowers sell. When borrowers sell, prices fall further. As prices fall, the real burden of nominal debt increases—your mortgage doesn't get smaller just because your house did. That higher real burden leads to more defaults and more forced selling, which pushes prices down again. The economy becomes a machine that converts falling prices into rising stress.

Debt-deflation explains why the aftermath can feel like pushing a boulder uphill with socks on. Even when interest rates fall, even when the initial panic ends, the legacy of forced selling and impaired balance sheets can keep the system fragile. It's not merely that people are pessimistic. It's that their net worth is lower, their debt is the same, and their ability to take risk has been impaired by math.

Restructuring is the catch-all term for what eventually breaks the logjam. Sometimes restructuring is formal: bankruptcy proceedings, debt exchanges, haircuts, negotiated settlements. Sometimes it's informal: maturity extensions, covenant waivers, "amend and extend" deals, quiet forbearance. Sometimes it's brutal and fast; sometimes it's slow enough to feel like a permanent condition. The purpose is the same: make obligations match reality again.

The speed of restructuring matters more than most people realize. There's a tempting fantasy that you can avoid pain by delaying it. In finance, delay often increases total pain. If losses are hidden and bad debts are rolled indefinitely, you can get "zombie" balance sheets—companies and banks that remain technically alive but too impaired to lend, invest, or grow. The economy can end up with a kind of financial sclerosis: resources stuck in unproductive uses because nobody wants to realize losses. The bill is paid not in one dramatic bankruptcy, but in a decade of sluggishness.

There's also a political economy to the aftermath. In the boom, leverage and risk-taking feel private; in the bust, they become public. People demand accountability, relief, punishment, protection. Regulators respond by tightening rules, increasing capital requirements, restricting certain forms of lending, forcing transparency, and redesigning plumbing that was previously ignored. These reforms can make the system safer. They can also shift activity to the parts of the financial world that are less regulated and less visible. The system learns—then it forgets—then it learns again, usually the hard way.

An important subtlety: deleveraging is rarely uniform. One sector may be paying down debt while another is taking it on. After a household credit boom collapses, households may become more conservative for years, while governments borrow more to stabilize the economy and cushion the shock. Meanwhile, some corporations may refinance cheaply and actually increase leverage, even as banks become cautious. The result is that "deleveraging" can be true in one part of the economy and false in another. The risk doesn't always disappear; it migrates.

That migration matters because it shapes what the next cycle looks like. If risk moves from banks to nonbanks, the next panic may not look like a classic bank run—it may look like a run on funds, a freeze in repo markets, a sudden evaporation of liquidity in a market that everyone assumed was deep. The aftermath, in other words, plants the seeds of the next architecture. Repair is real, but so is adaptation.

Recovery stories are often told with a single shape: a V. Down hard, up fast. It's comforting, like a movie where the hero gets knocked down and then wins in the final act. But debt crises are frequently messier. A boom driven by leverage tends to leave behind balance sheets that take time to heal. That healing can produce an L-shaped experience: the crash is sharp, the rebound is partial, and then growth drags because so much cash flow is devoted to repayment rather than expansion.

This doesn't mean the economy is doomed after every bust. It means the timeline is different. If the bust is a liquidity event—panic but not deep insolvency—recovery can be quick. If the bust reveals widespread overindebtedness, recovery is often slow because you can't sprint while carrying a heavy pack. Debt has to be reduced, written down, or grown out of. Those are not fast processes.

For investors, the aftermath can be psychologically treacherous because it punishes the two instincts that feel most natural after a crash: the instinct to swear off risk forever, and the instinct to make all your money back immediately. The first can leave you permanently underinvested in a world where assets eventually recover and compound. The second can lure you into exactly the kind of leverage and desperation trades that panics are designed to destroy. Post-crash markets can be generous to patience and merciless to urgency.

For households, the aftermath is where financial habits are rewritten. The boom rewarded convenience: low payments, easy refinancing, "you deserve it." The bust rewards resilience: emergency funds, fixed rates, smaller obligations, the ability to absorb a hit without begging a lender for mercy. The most valuable asset in the aftermath is not a hot tip. It's breathing room.

One of the clearest signs that an economy is truly healing is not a stock index hitting a new high. It's boring behavior returning in a healthy way: lending based on income rather than collateral fantasies, companies investing because they see demand rather than because credit is cheap, households saving because it's prudent rather than because they're terrified. The goal is not to eliminate risk—an economy without risk is an economy without growth—but to rebuild a system where risk-taking is financed in a way that can survive disappointment.

There's a final irony worth holding onto, because it keeps you honest. The aftermath feels like an overreaction while you're living through it. People become conservative. Rules get stricter. Everyone is suddenly a skeptic. And then, slowly, as the memories fade and the good years stack up, that conservatism starts to feel unnecessary. The guardrails begin to look like obstacles. The old temptations return in new packaging. The cycle doesn't repeat because humans are stupid. It repeats because humans are human: we heal, we forget pain, and we start believing again.

Actionable takeaway: Build your personal "post-bust protocol" before the next bust arrives. Concretely, choose one balance-sheet repair move that reduces your vulnerability to deleveraging spirals:

- (i) move any variable-rate, refinance-dependent debt toward fixed payments;
- (ii) build a cash buffer that covers at least one month of essentials, then expand it toward three;
- (iii) reduce any investment borrowing (margin, HELOC investing) to zero unless you could repay it quickly without selling assets;
- (iv) set a rule that major purchases must be affordable even if income drops 10% for a year.

Chapter 4

Echoes of the Past: Centuries of Speculation

They say history doesn't repeat itself, but in finance, it certainly rhymes. Before we can understand modern crypto-crashes or housing bubbles, we must look back at their ancestors. From Dutch burghers trading flower bulbs for the price of mansions to the railway tycoons who laid tracks to nowhere, this chapter explores the granddaddies of speculation. We will see that while the technology changes-from wooden ships to steam engines to ticker tape-the human emotions driving the market remain stubbornly the same.

4.1 Reading the Past: How to Study Financial History

If you've ever watched old footage of traders in 1929—ties loosened, paper flying, faces tight—you've probably felt a quiet, smug thought slide in: *How could they not see it?* The chart looks like a ramp into thin air. The headlines, when you line them up afterward, read like a checklist of warning signs. In hindsight, the crash seems less like an ambush and more like a slow-motion confession.

That feeling is the first trap. Financial history punishes the reader who walks in like a judge. It rewards the reader who walks in like an investigator.

An investigator doesn't start with the verdict. An investigator starts with the scene. Who had what information? What rules governed their choices? What counted as "safe" in their world? Which risks were obvious, and which were invisible because the institutions to measure them didn't exist yet? The goal isn't to admire or condemn. The goal is to reconstruct the mental weather of the time: what people could reasonably believe on an ordinary Tuesday, before the extraordinary Wednesday arrived.

There's a practical reason to care about this mindset: it changes what you extract from the past. If you read history as a parade of idiots and villains, you'll get moral satisfaction and almost no usable pattern recognition. If you read it as a sequence of constrained decisions under uncertainty, you start noticing the repeatable machinery—the ways stories turn into prices, the ways leverage turns prices into fragility, and the ways fragile systems break when confidence gets a pinhole.

The most expensive bias: mistaking inevitability for insight

The cleanest lie financial history tells is the lie of inevitability. Once a crash has happened, every earlier moment gets re-lit by the flash. A newspaper column that seemed mildly worried becomes a prophecy. A politician's speech becomes a smoking gun. A skeptical investor becomes a hero. The narrative tightens into a neat rope: up, up, up, then down—as if gravity was doing the trading.

Real time does not feel like that. Real time is fog, not rope.

Markets, especially during booms, are noisy machines that reward the confident and punish the cautious in small, daily ways. If you were an investor in a mania, you didn't wake up to a banner reading *Bubble Day 387*. You woke up to friends getting richer, newspapers praising innovation, respectable bankers explaining why the old rules no longer applied, and a constant, low-grade fear of being left behind. That combination produces a particularly dangerous form of intelligence: the ability to build a sophisticated argument for whatever price is currently on the screen.

Hindsight bias doesn't just make you unfair to people in the past. It makes you reckless in the present, because it convinces you that the next crash will also look obvious—to you, of course—right before it happens. In reality, if you can see a collapse clearly enough to bet on it comfortably, so can everyone else, and the price will already reflect the fear. The past looks crisp because the outcome is known. Your future will not grant you that favor.

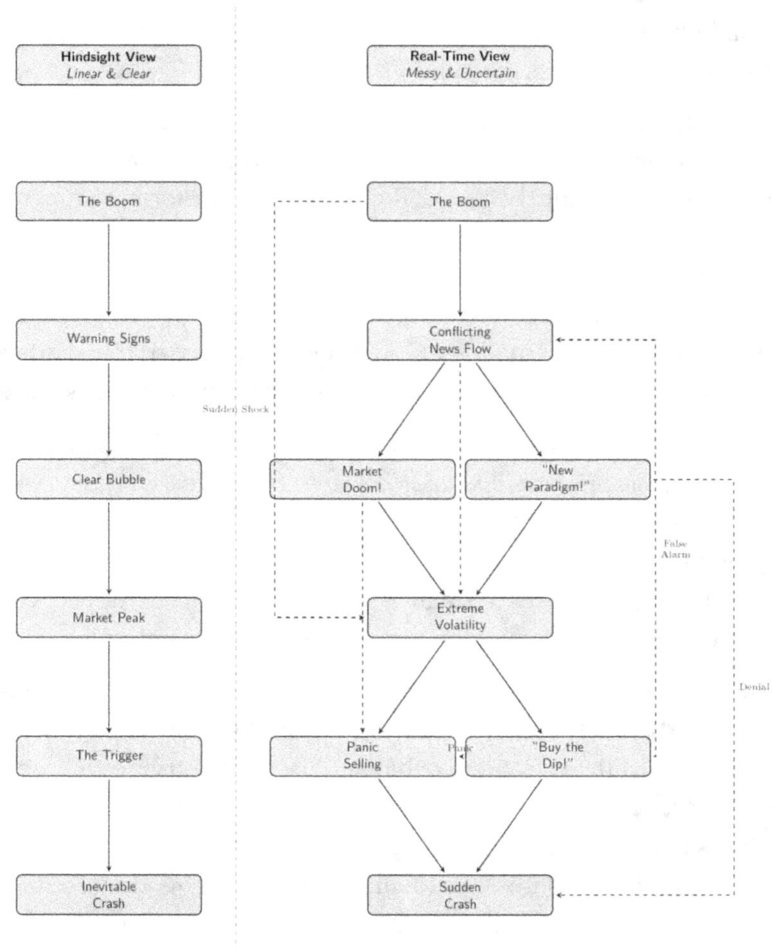

A useful discipline is to force yourself to write two timelines for any episode. One is the *outcome timeline*: the clean story you tell after the fact. The other is the *information timeline*: what participants actually knew, when they knew it, and what they could plausibly infer. The second timeline is usually uglier, full of false alarms and partial recoveries. It's also the one that teaches you how bubbles persist. They don't levitate on pure delusion; they surf on ambiguity.

The ``it was irrational'' shortcut (and why it fails)

Calling past investors irrational is a convenient way to stop thinking. It's also a way to miss the point.

In nearly every boom-bust episode, you can find rational components tangled together with social imitation, career pressure, and genuine technological change. Railways did transform commerce. The internet did rewrite business models. Homes do provide shelter and utility. Even tulips, mocked as pure folly, sat inside a real culture of collecting, status, and specialized markets. When you dismiss a mania as nothing but madness, you throw away the part that matters most for investors: *how sensible ideas get stretched into dangerous prices.*

A better question than "Were they irrational?" is: *What did they believe would happen next, and what had to be true for those beliefs to pay off?* Once you phrase it that way, you can map a boom as a stack of assumptions. Some assumptions are reasonable. The bubble forms when the price requires too many assumptions to all be true at once—and when leverage makes the cost of being wrong explosive.

There is also a darker, more human reason the "irrational" label fails: many participants weren't behaving like philosophers maximizing long-run wealth. They were behaving like people trying to keep a job, protect a reputation, or avoid humiliation. A bank manager who refuses to join the lending party might be right in ten years and unemployed in ten months. "Prudent" can be a career-ending strategy when everyone is getting promoted for being bold. Crowds don't just spread optimism; they enforce it.

Context is not trivia; it is the operating system

Studying a past market without its legal and institutional context is like studying a modern smartphone by analyzing the glass. You'll miss the operating system. The laws of contract, the structure of banking, the availability of credit, the speed of communication, and the credibility of government backstops—these are not background details. They are the machine.

A panic in a world without deposit insurance behaves differently from a panic in a world with it. A stock boom under a gold standard behaves differently from a boom under an elastic fiat-money regime. A market where shares are bought on margin with a handshake behaves differently from a market with formal collateral rules and daily mark-to-market. The *same* human emotions can generate very different outcomes because the institutional plumbing changes how stress propagates.

When you read about a nineteenth-century bank run, you're not watching a quaint antique version of the same thing you've seen in movies. You're watching a particular technology of finance: banks funding long-term, illiquid assets (like loans) with short-term, withdrawable liabilities (deposits). That mismatch is not a moral failing; it's a business model. It works beautifully until confidence wobbles. Then it becomes a race: the first depositor gets cash, the last depositor gets a promise.

Context also determines what counted as "information." Today, a single rumor can travel globally in minutes. In earlier centuries, news moved at the speed of horses, ships, and gossip. That doesn't mean markets were calmer; it means uncertainty lasted longer. Prices could drift on speculation not because people were foolish, but because they had no reliable way to *resolve* disagreement quickly. When you can't settle what's true, you trade stories.

So one of the investigator's first moves is almost boring, but transformative: before you study the chart, study the constraints. Who could borrow? Against what collateral? Under what enforcement? With what penalties for default? And from whom?

Beware the survivor's museum: what gets preserved is not what mattered most

Financial history is built from records, and records are biased. Some companies keep archives; others burn down, merge, fail, or vanish. Some governments track statistics carefully; others don't have the administrative capacity. Newspapers survive; private letters don't. That means the past you can read is not the past as it happened. It's the past that left paperwork.

This produces a subtle distortion: we over-learn from winners and under-learn from the countless quiet failures that never made it into a headline. In a mania, the spectacular successes become the story. The bankruptcies become footnotes. Later, the footnotes disappear, and you're left with a museum of survivor artifacts: the famous firms, the famous speculators, the famous scandals. The system that created the boom—the lending standards, the incentive structures, the widespread small decisions—gets less attention precisely because it is less narratively satisfying.

An investigator compensates by asking: *What am I not seeing because it didn't survive?* If you're reading a history of a bubble built around a handful of celebrated tycoons, look for the uncelebrated infrastructure: the brokers who extended credit, the accountants who blessed balance sheets, the local banks who concentrated exposure, the lawmakers who created loopholes. When a crisis hits, the failures that matter most are often the boring ones.

A practical technique is to look for aggregate indicators that are harder to romanticize: measures of credit growth, interest rates, bank reserves, defaults, or insolvencies. These numbers are not perfect (they come with their own measurement problems), but they can counterbalance the gravitational pull of anecdote. Stories tell you what it felt like. Aggregates tell you how widespread it was.

Data is not neutral: series breaks, changing definitions, and the mirage of precision

Historical finance has a seductive aesthetic: long time series, old charts, centuries of prices marching across the page like a disciplined army. It feels scientific. It can also be an optical illusion.

A "dollar" is not a timeless unit. A "bank" is not a timeless institution. A "stock market" is not a timeless venue. When you compare financial data across eras, you're often comparing objects that share a name but not a meaning. Definitions shift. Borders move. Accounting standards change. The set of traded securities evolves. Even the *purpose* of markets changes—from financing wars, to financing canals, to financing railroads, to financing corporate empires.

These changes create what researchers call *series breaks*: points where a data series looks continuous but the measurement method or underlying concept has changed. If you don't notice the break, you might attribute a sudden jump to exuberance when it was really a change in reporting. Or you might mistake stability for safety when the data simply stopped capturing the riskiest corner of the market.

Then there's inflation, the quiet saboteur of naive comparisons. A price chart without purchasing-power context is a story with missing chapters. A stock index that "returned" 6% a year over a century might hide long stretches where the real return (after inflation) was negative. Conversely, a nominal crash might look catastrophic until you realize the preceding inflation had already eroded real values. If you want to feel how people experienced wealth, you need to translate nominal numbers into lived buying power: wages, rent, bread, coal, land.

And even when data is clean, it can still mislead through false precision. A chart with a crisp line implies accuracy down to the pixel. Historical prices often come from thin markets, irregular trading, or recorded transactions that were not representative. Sometimes the "price" is the last trade of a week, or a quote in a newspaper that may have been optimistic advertising. The point is not to distrust all historical data. The point is to treat it with the same skepticism you'd apply to a company press release during a boom: useful, but not holy.

The danger of moral fables: villains, heroes, and the comfort of a simple cause

Popular financial history loves a villain. Preferably a charismatic one. Ideally one who later confesses.

This creates a second shortcut: the idea that a crash has a single cause, a single spark, and a single person (or small cabal) who lit it. In reality, spectacular wrongdoing often exists alongside a more important truth: booms build system-wide fragility. Fraud can accelerate the fall, but it usually doesn't create the underlying instability by itself.

If you want to understand contagion, don't stop at greed. Greed is constant across time. What changes is the mechanism that allows greed to scale.

The scalable mechanism is typically credit. Credit is a tool that turns optimism into purchasing power. When credit expands easily, prices can rise far beyond what cash buyers could support. That rise becomes proof of the story, drawing in more participants, which invites lenders to loosen terms, which pushes prices higher. The boom starts to feed itself. Then something interrupts the loop: interest rates rise, collateral values wobble, a key institution fails, or the story loses its charm. Credit tightens, forced selling begins, and the same feedback loop runs in reverse.

That's why an investigator treats moral narratives as clues, not conclusions. Corruption matters, but it's rarely sufficient. Innovation matters, but it's rarely safe by itself. Psychology matters, but it always expresses through institutions. The past becomes most useful when you can see multiple causes interacting, rather than hunting for a single dramatic one.

Thinking in mechanisms: turning episodes into a repeatable template

To study financial history for practical use, you need a translation tool: a way to turn a colorful episode into a mechanism you can recognize in new clothing.

One effective template is to describe each boom-bust as a sequence of states:

- A *displacement*: a shock or change that makes people believe the future has become richer (a new technology, new trade, new policy, a war ending, a new source of credit).
- *Credit expansion*: the financial system finds ways to lend against the new story (new instruments, easier terms, new intermediaries).
- *Euphoria*: rising prices become evidence; skepticism becomes expensive; participation broadens.
- *Distress*: the first cracks appear; a few institutions stumble; prices wobble.
- *Revulsion*: confidence breaks; forced selling cascades; liquidity evaporates.

The point of a template isn't to reduce history to a formula. It's to prevent yourself from being hypnotized by the costumes. Tulips, railways, dot-coms, condos, cryptocurrencies—the object changes, but the loop of narrative, credit, and crowd behavior rhymes in ways that are uncomfortably consistent.

Mechanisms also help you avoid another bias: the belief that every crash must look like the last famous one. Many crises are not stock-market crashes; they're banking crises, currency crises, or slow-motion debt deflations. Some arrive as sudden cliffs. Others arrive as a long, grim staircase. If your mental model is only "market goes up, market crashes," you'll miss the subtler dynamics: funding markets freezing, collateral haircuts widening, credit spreads gapping out, or depositors quietly leaving before the line forms.

The investigator's question is always: *Where is the fragility? What happens if confidence drops 10%? Who is forced to sell? Who cannot refinance? What relies on rolling short-term funding?* History is a catalog of answers.

A simple field method: reconstruct the trade, not the tale

When you read a historical episode, try a practical exercise: reconstruct the representative trade that the crowd was doing at the top. Not a metaphorical trade. An actual trade.

Who was buying? With what money? How much was borrowed? At what rate? What collateral backed the loan? What income stream was expected to service the debt? What price increase was implicitly required for the trade to be attractive? What would make the trade fail?

This approach cuts through romance. It forces you to translate excitement into arithmetic.

For example, if a boom is funded by short-term borrowing against long-term assets, then refinancing risk is the Achilles' heel. If a boom depends on ever-rising prices to keep collateral sufficient, then even a small price decline can trigger margin calls and forced selling. If a boom depends on a new technology delivering profits quickly, then delays—not failure, just delays—can be lethal.

Reconstructing the trade also clarifies why bubbles can persist longer than skeptics think. Many trades work *as long as* financing remains available and prices don't fall too far too fast. That "as long as" is the hidden engine of complacency. It is entirely possible for intelligent people to participate in a trade they privately suspect is fragile, because their true bet is on the *continuation of funding*, not the permanent truth of the story. In other words: they are not investing; they are timing.

History becomes much less mysterious once you notice how often the crowd is timing.

How to read old opinions without being fooled by them

Historical markets leave behind a trail of opinions: pamphlets, speeches, editorials, diaries, testimony. These sources are gold, because they let you hear the uncertainty in real time. They're also dangerous, because they're persuasive.

The investigator treats opinions as data with incentives attached.

A promoter selling shares will sound confident. A banker seeking deposits will sound safe. A politician defending a policy will sound righteous. A critic who was ignored will sound prescient after the crash. None of that tells you who was right; it tells you what they needed others to believe.

A useful trick is to classify contemporary voices into three buckets:

- *Participants* (they profit if the boom continues),
- *Intermediaries* (they profit from activity regardless of direction, like brokers and underwriters),
- *Outsiders* (they profit from being correct later, like some journalists, academics, or rivals).

Then ask: *What was the cost of being wrong for each group?* The higher the cost, the more likely the voice is shaped by motivated reasoning. This doesn't mean participants are lying; it means they are often convincing themselves first. Human beings are remarkably good at turning self-interest into sincere conviction, especially when the crowd applauds.

Actionable takeaway: build a ``historian's checklist" for your own investments

Studying financial history pays off when it changes how you look at your *current* portfolio. Here's a concrete habit that turns the investigator's mindset into something you can use immediately:

Once a quarter, pick one asset you own (or strongly want to buy) and write a one-page "history-style" brief about it. Include:

- The dominant narrative in one sentence (what everyone thinks is true).
- The financing channel (who is lending, on what terms, and what collateral supports the system).
- The fragility point (what has to stay smooth for the trade to keep working).
- The missing data (what you'd want to know but can't easily observe).
- Two alternative endings: one benign (the story is mostly true), one ugly (the story stumbles).

If you can't fill in the financing channel or the fragility point, you're not doing analysis; you're doing vibes. History's main gift is not predicting the next crash date. It's teaching you where to look for the cracks *before* the chart makes them obvious.

4.2 The Dawn of Speculation: Tulips and the South Sea

The cliché version of Tulip Mania goes like this: sensible Dutch citizens traded houses for flowers, then hurled themselves into canals when the petals fell out of fashion. It's a perfect morality play—quick, colorful, and wrong in the way good myths often are. The real story is less cartoonish and more useful, because it shows something that still drives markets today: when a new financial *method* spreads faster than the social rules that keep promises enforceable, prices don't need to be "crazy" to become unstable.

Amsterdam in the 1630s was a city built for trade: narrow houses, wide ledgers, and a culture that treated credit as a tool rather than a sin. Tulips arrived as an exotic luxury with a collector's appeal, the way rare watches or limited-edition sneakers work now-objects that are partly about beauty and partly about membership. Certain bulbs, especially those with dramatic "broken" flames of color, were genuinely scarce. Scarcity is not irrational. Scarcity is just a fact that becomes a story once people start comparing notes.

The part that matters isn't that people liked flowers. It's that people learned how to trade *future delivery* of those flowers.

A tulip bulb spends much of the year underground. You can't simply hand it over every afternoon the way you can hand over a share certificate. That physical inconvenience nudged the market toward contracts: agreements written in taverns and meeting rooms to buy or sell a bulb at a later date, often after lifting season. In modern language, that's a futures-style trade: you're not buying the object; you're buying a commitment about the object. Commitments are light and fast. They can multiply far more quickly than bulbs can.

Here's the first echo you'll hear again and again across centuries: once a market starts trading promises, it can expand without waiting for the real-world thing to expand. Prices can move on expectations alone. That's not necessarily fraud—it's finance. But it changes the physics. A cash market is limited by cash. A promise market is limited by confidence.

What, exactly, was being traded in the peak months? Often not a bulb sitting in a box, but a claim: "I will deliver this variety later; you will pay that price then." These agreements could be resold, which made them feel a lot like assets. When you can flip a contract to someone else, you stop thinking about taking delivery. The trade turns into a game of *exiting*.

And when a trade is about exiting, social mood becomes a pricing input.

At this point, the legend usually barges in: "One bulb cost as much as a house." Sometimes the comparison is exaggerated; sometimes it's based on rare varieties; sometimes it's a rhetorical flourish that hardened into "fact" through repetition. But it contains a kernel worth respecting: at the peak, certain named bulbs were quoted at levels that dwarfed ordinary wages. Even if the number varies by source, the *distance* between the luxury object and the everyday paycheck is the real signal. It tells you who was in the room (people with surplus wealth), and it tells you what kind of emotion was driving the trade (status, urgency, bragging rights, fear of missing out).

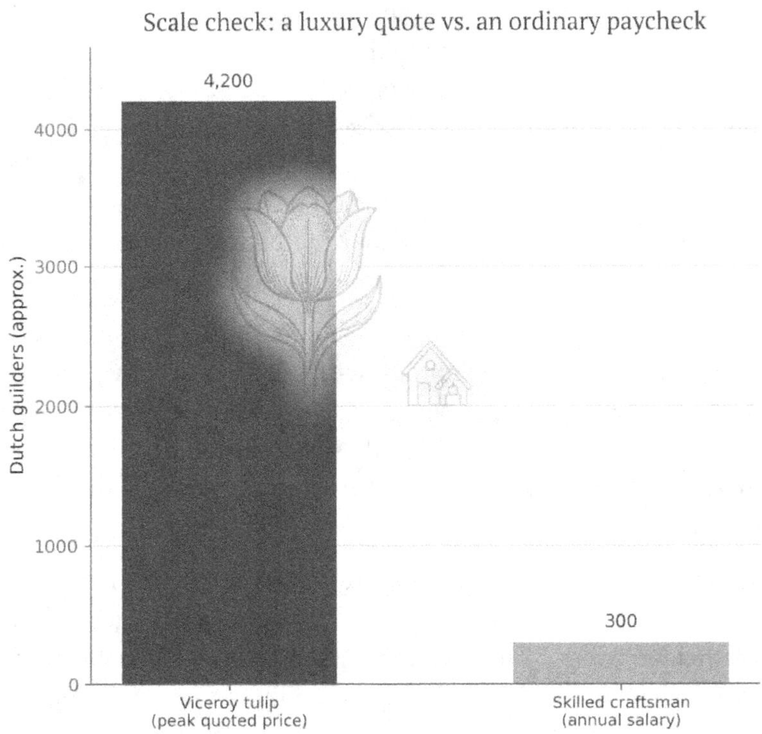

Illustrative comparison: quoted peak prices and wages vary across records; the point is the order of magnitude.

Once you see the gap in that chart, the next question isn't "How could they?" The answer is a mix of novelty and social technology. These were not anonymous, regulated exchanges with standardized margin requirements and courts designed to churn through disputes at scale. Many agreements leaned on reputation, local norms, and the threat of being shunned. That works wonderfully—until it doesn't.

The crack in the tulip market wasn't just a price decline. It was a problem of *enforcement*. When prices turned, some buyers refused to honor contracts struck at peak levels. If the contract is informal, the cost of walking away might be mostly social. That's manageable when the crowd is euphoric and participation is narrow. It becomes corrosive when the mood flips and everyone is looking for a reason they "shouldn't have to pay." The damage shows up less as a single dramatic bankruptcy wave and more as a sudden distrust: if I can't trust your signature, why would I trade with you tomorrow?

That's a form of contagion that doesn't need banks. It spreads through belief in the basic rules of the game.

Tulip Mania also deserves a second, more sympathetic read. Not because the prices were "reasonable," but because the participants were operating at the edge of a new financial idea: that you could slice time into tradable pieces. Modern markets are built on that idea. If you've ever bought a stock because you believe you can sell it later to someone else at a higher price, you've participated in the same underlying logic. The difference is infrastructure: exchanges, clearing, regulation, standardized contracts, and deep liquidity that can absorb panic (most days). The Dutch were experimenting in a world where the rails were still being laid.

Now move forward less than a century and you'll find the second foundational tale-one that doesn't revolve around flowers or fashion, but around the most serious thing a government can offer: its own debt.

London in 1720 had coffeehouses that functioned the way trading apps and social media do today: places where rumor, aspiration, and "hot tips" mixed into an intoxicating brew. The South Sea Company wasn't selling petals; it was selling a financial conversion. The basic scheme, stripped of period jargon, was a political-financial bargain: the company would assume chunks of Britain's national debt, and in return it would receive favorable terms and privileges. Debt-normally dull, slow, and respectable-was turned into something with a story and a ticker.

And once government finance becomes a story, speculation gains a powerful ally: legitimacy.

The South Sea pitch had several ingredients that make bubbles particularly potent. First, it wore a suit. It wasn't a back-alley wager; it was Parliament-adjacent, discussed by elites, and tangled up with the machinery of the state. Second, it involved a kind of alchemy that investors love: turning a messy pile of obligations into a clean, marketable asset. Third, it carried the perfume of empire-trade prospects in distant seas that sounded limitless to people who couldn't verify them easily.

When the price began rising, the rise itself became a political fact. A soaring share price makes supporters look smart, makes critics look petty, and creates constituencies who want the rise to continue. That feedback loop is one reason state-linked bubbles are so dangerous: they can recruit not only private greed but public pride.

The speed of the South Sea run-up—and the speed of its collapse—reveals another recurring mechanism: financial contagion thrives on imitation. The most obvious imitation was the eruption of copycat ventures, companies with grand names and vague purposes, designed less to build something than to ride the wave of investor appetite. The less obvious imitation happened inside people's heads. Once your neighbor doubles their money on a stock tied to national destiny, caution starts to feel like a personality defect.

Then the fragile parts of the structure mattered. A boom fueled by leverage (borrowed money) is a boom with a hidden trapdoor, but even without widespread margin lending, a boom fueled by *concentrated belief* is still brittle. When everyone is on one side of the boat, a small tilt feels like a storm. When the South Sea price began to wobble, confidence didn't simply "cool." It reversed. Promises that were easy to sell on the way up became hard to sell on the way down. Liquidity—the ability to turn an asset into cash quickly without taking a brutal discount—vanished precisely when it was most needed.

The aftermath is often misunderstood. It's tempting to treat the crash as a morality ending: the market sinned, got punished, and learned its lesson forever. Markets do not learn forever. They learn briefly, then hire new employees. What did change, though, were institutional reflexes: a sharper awareness that speculation could become a public issue, not merely a private embarrassment. Political scandal and investigations followed. Reputations shattered. Rules tightened in some areas, sometimes clumsily. And yet the broader system did not return to a pre-financial state. Capital still sought outlets. Investors still chased return. The machinery kept evolving.

That's the most modern part of these early stories: a crash rarely kills the appetite for speculation. It changes the *wrapper*.

Tulips and the South Sea are often presented as opposites: one frivolous, one serious; one about fashion, one about statecraft. Look closer and they rhyme in three crucial ways.

First, both episodes involved a gap between the real asset and the tradable claim. Tulips became contracts. National debt became a convertible, marketable story. When claims become easier to trade than the underlying reality is to evaluate, prices can detach from what cautious, cash-based buyers would pay.

Second, both depended on trust structures that were stressed by scale. In Amsterdam, trust was social and local; in London, trust was political and institutional. In both cases, when the crowd grew and the stakes rose, the trust system started creaking. Once the creak is audible, people begin trading not on value but on the anticipated behavior of everyone else. That's the pivot point where markets start acting like crowds.

Third, both produced contagion. Not just financial loss, but social imitation—an epidemic of participation driven by observation rather than analysis. The first person in makes money; the tenth feels smart; the hundredth feels safe; the thousandth feels *normal*. When "normal" is set by a rising price, falling prices feel like a violation of reality itself. That emotional whiplash is a key accelerant of crashes.

There's a final lesson hidden beneath the powdered wigs and canal-side taverns: financial innovation is not a villain. It's a power tool. Futures-style contracts and debt conversions can be useful, even essential. The danger is timing. When the tool spreads before norms, enforcement, and risk management catch up, it creates a window where optimistic stories can be financed at scale and then unwind chaotically.

That window is where manias thrive. It's also where serious investors get lulled into treating novelty as diversification. New instruments don't automatically reduce risk; sometimes they concentrate it by creating the illusion that risk has been "distributed," when in fact it has merely been repackaged.

Actionable takeaway: when the story is new, interrogate the promises

The practical habit to steal from these early bubbles is simple enough to do with a notebook and stubbornness:

Whenever you see a hot market built around something "new," write down the exact promise being traded—and ask who enforces it when prices fall.

Not the marketing promise ("this will change everything"), the contractual promise:

- What do buyers actually own: the asset, a claim on future delivery, a share of profits, a token, a right to convert, a right to refinance?
- What makes the promise credible: courts, clearinghouses, collateral, regulation, reputation, or politics?
- What happens if a large fraction of participants decide, all at once, that the promise is inconvenient?

If the answer relies mostly on mood, honor, or the assumption that the crowd will remain cheerful, you're not looking at a timeless opportunity. You're looking at a trust structure under stress—the same kind that snapped in Amsterdam and London, long before markets had skyscrapers to fall from.

4.3 Industrial Dreams and Banking Panics

A tulip bulb can be precious, but it can't carry freight. A share certificate can soar, but it can't shorten a journey. Railways were different. They were loud, physical, and persuasive in the way only a new machine can be: iron tracks cutting through fields, stations rising like small cathedrals, timetables turning distance into something you could schedule. If you wanted a nineteenth-century investor to loosen their grip on caution, you didn't need poetry. You just needed to put them on a train.

That's why the Victorian era is the perfect laboratory for speculation. The dream was real. The technology was revolutionary. And the money, once it started pouring in, didn't just misprice an object—it misallocated a nation's savings into projects that could take decades to justify, if they ever did.

The Railway Mania of the 1840s in Britain wasn't built on nonsense; it was built on a truth that was easy to overextend: transport transforms economies. A new line could turn a sleepy town into a commercial hub, slash shipping costs, and compress time in a way that made old business models look medieval overnight. Investors weren't hallucinating those benefits. They were paying for them—sometimes sensibly, often twice, and occasionally ten times.

The problem with real revolutions is that they invite fake certainty. Once you've seen a locomotive outrun a horse, it becomes tempting to believe that every railway plan is the future, every projected dividend is inevitable, and every skeptic is merely slow. The future stops being a probability distribution and becomes a moral identity: you are either *for progress* or you are an obstacle.

Railway shares were tailor-made to pull in a crowd because they combined two intoxicating elements: a story you could touch and a financial structure that made participation feel easy. Many offerings used *part-paid shares*-you didn't have to hand over the full amount up front. You paid an initial deposit and promised to pay the rest later if the company issued a "call." In plain English, it was a down payment on a dream, with a potentially nasty invoice arriving in the future.

This structure did something subtle to the mind. It shrank the pain of buying and pushed the pain of owning into an uncertain tomorrow. It also turned the market into a confidence machine: as long as prices rose, investors felt richer, which made future calls feel manageable, which encouraged more buying, which pushed prices higher. When prices fell, the logic inverted with equal force. A call on an underwater share feels less like an investment request and more like a punishment.

If the tulip market showed how fast *promises* can circulate, railways showed how fast *capital* can be committed to long-lived projects. Tracks, bridges, tunnels-these weren't assets you could quietly unwind when sentiment soured. Once the money went into earth and stone, it stayed there, indifferent to market moods. That mismatch—short-term enthusiasm funding long-term construction—created a chronic vulnerability: investors could change their minds overnight, but railways could not.

Another feature of railway booms is that they're strangely democratic. Not because everyone gets rich, but because everyone feels invited. The romance of a national network makes speculation feel civic. Newspapers print routes. Towns lobby for stations. Politicians pose as patrons of progress. A railway share becomes more than a claim on profits; it becomes a vote for modernity. When investing is wrapped in identity, price discipline weakens. Selling doesn't feel like prudence; it feels like betrayal.

And then the bills arrive. Construction costs overrun, as they love to do. Competition appears, as it always does when profits are rumored to be easy. Lines are built in duplicate-two tracks to the same thinly populated area because promoters could raise money on forecasts, not on passengers. The air leaks out not necessarily because the revolution was fake, but because the *pricing* assumed a revolution without friction: no delays, no recessions, no politics, no bad weather, no engineering surprises, no rivals. Real life is mostly those things.

Now comes the darker Victorian lesson: when industrial dreams collide with a fragile banking system, a speculative hangover can become a panic.

In the late nineteenth and early twentieth centuries, the United States in particular lived through a series of financial panics—1873, 1893, 1907—episodes that feel like different verses of the same harsh song. The common melody was a banking structure that could amplify fear into catastrophe. Banks took deposits payable on demand and turned them into longer-term loans and investments. That's not villainy; it's how banking works. But it means confidence is not a nice-to-have. It is a core input. When it vanishes, the business model stops functioning.

The Panic of 1873 is a useful example because it ties the industrial dream directly to financial plumbing. Railroads weren't only built with equity from cheerful shareholders; they were built with bonds—debt issued to investors who wanted steady interest. Those bonds were often marketed and underwritten by powerful intermediaries. When a major firm tied to that railroad finance failed, it didn't just hurt its own clients. It raised a terrifying question for everyone else: *Who's next?*

In a modern crisis, the most important fights often happen behind the scenes—in repo markets, money market funds, interbank lending. In 1873, the scene was more visible and more physical: anxious depositors, shuttered institutions, sudden halts in trading, employers unable to make payroll because credit snapped like a dry branch. A bank run is not a philosophical debate about value. It is a sprint for liquidity.

Liquidity is a plain concept with brutal consequences: the ability to get cash when you need it. A bank can be solvent in a slow, accounting sense-its assets may exceed its liabilities-yet still fail because it cannot turn those assets into cash quickly enough to meet withdrawals. Loans can be "good" and still be illiquid. Railroad bonds can be "sound" and still be unsellable in a panic. In crises, timing outranks truth.

The pre-Federal Reserve era made this timing problem worse. There was no central, automatic backstop to lend freely against good collateral and calm the stampede. Instead, the system relied on a patchwork of private coordination and improvisation. Clearinghouses—associations of banks—sometimes acted like proto-central banks, pooling resources, sharing information, and attempting to prevent the weakest link from snapping first. Their efforts could be heroic. They could also be too little, too late, or politically constrained by who trusted whom.

The Panic of 1907 is the episode that makes the mechanics easiest to picture. A shock hits a corner of the financial world. Rumors spread faster than balance sheets can be checked. Depositors and short-term lenders decide they'd rather hold cash than promises. Institutions that relied on rolling over funding discover that "roll" has become "stop." Assets that were comfortably priced yesterday have to be sold today—quickly—because cash is leaving.

Forced selling is the crisis accelerant that turns fear into price collapse. You don't sell because you think the asset is worthless; you sell because you need cash *now*. A wave of forced selling pushes prices down, which makes collateral less valuable, which triggers more demands for cash, which forces more selling. The market stops behaving like a weighing machine and starts behaving like a trap door.

That feedback loop—confidence cracking into withdrawals, withdrawals into liquidity stress, liquidity stress into fire sales—is so central to financial history that it's worth seeing as a system, not as a sequence of anecdotes.

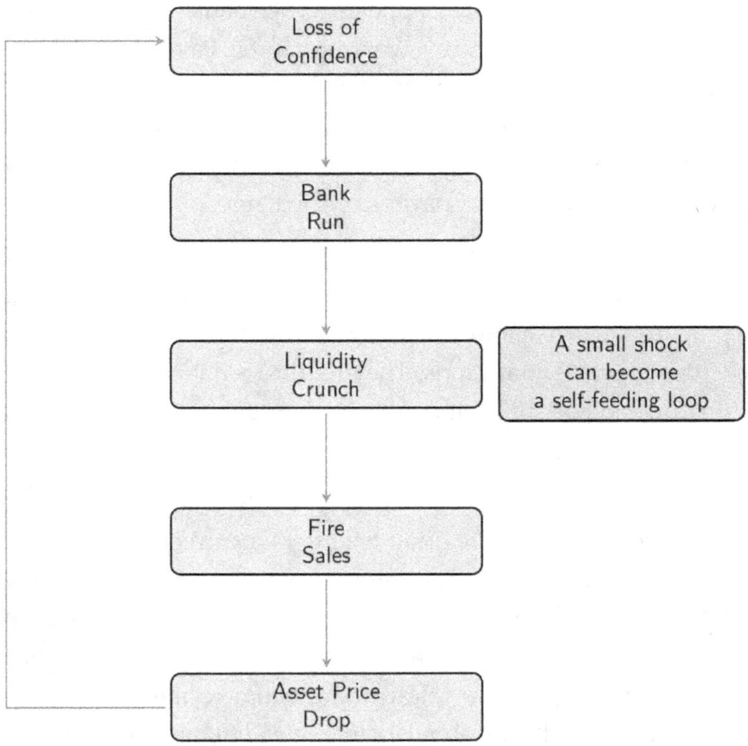

Once you see the loop, the Victorian-era panics stop feeling like random thunderstorms. They become the predictable weather of a system built without enough shock absorbers.

The most painful twist is that the loop can punish even "good" behavior. Imagine a bank manager in 1907 holding a portfolio of loans that will mostly pay back over time. If depositors demand cash today, the manager must find cash today. That usually means selling whatever can be sold—often the safest, most marketable assets first—because those are the only ones buyers will touch in a panic. The bank's balance sheet deteriorates not because it took the wildest risks, but because the crisis forces it to liquidate its best assets at the worst moment. Prudence becomes fuel for the fire.

The same dynamic hits ordinary businesses. A manufacturer might have real orders, real customers, and a real future, yet still collapse because short-term credit disappears. When banks hoard cash to survive runs, they stop extending working capital. Payrolls bounce. Suppliers demand upfront payment. Commerce seizes in places far removed from the original spark. This is one reason financial crises feel unfair: they spread through the credit system like smoke through vents, choking rooms that never saw the flame.

Railway booms and banking panics are often told as separate tales-one about speculative exuberance, one about institutional fragility. In practice they are entangled. Long-gestation projects are especially sensitive to credit conditions. They require refinancing, ongoing investment, and patient capital. A panic shortens everyone's time horizon. When time horizons shrink, projects that are viable in a calm world become impossible in a frightened one.

This is also where the Victorian era offers a corrective to a modern misconception: that bubbles are only about ridiculous assets and gullible people. Railways were transformative. Many lines eventually earned steady returns. The tragedy is that a genuine technological shift can still produce a destructive boom because finance is not merely about truth; it's about *timing*, *liquidity*, and *coordination*. A great invention does not immunize investors from bad entry prices, overbuilding, or a funding model that collapses under stress.

It's tempting to search for the single villain in these episodes: the promoter, the crooked director, the reckless lender. Sometimes such figures exist in abundance. But the deeper villain is structural: a system that allows optimism to scale quickly and then provides no orderly way to slow it down. In a world without a credible lender of last resort, without consistent rules for emergency liquidity, and with fragmented banking networks, fear becomes a solvency event.

Reforms after these panics weren't acts of philosophical enlightenment; they were acts of exhaustion. Repeated crises taught a practical lesson: if the system is going to fund long-term growth with short-term claims, it needs mechanisms to prevent every tremor from becoming a stampede. Emergency currency measures, clearinghouse coordination, and eventually central banking were attempts to put a circuit breaker into the loop you saw in the diagram.

Yet even with better plumbing, the emotional core remains stubbornly familiar. Confidence is still the most underpriced asset in finance—ignored when abundant, priceless when scarce. Liquidity is still treated like air: nobody thanks it on normal days, and everyone panics when it disappears. And technological revolutions still invite investors to confuse "the future is big" with "every price is justified."

Actionable takeaway: stress-test the funding, not just the idea

A Victorian-era habit that still works is to separate the *dream* from the *funding*. Before buying into any fast-growing theme—AI, clean energy, biotech, private credit, real estate—run a quick personal audit that focuses on the panic loop:

- Identify the nearest equivalent of a bank run: *Who can demand cash back quickly?* (depositors, short-term lenders, ETF holders, money market investors, venture LPs, refinancing markets).

- Identify the liquidity pinch point: *What has to be sold or rolled over if confidence drops?*
- Identify the forced-sale asset: *What gets dumped first, and how would that price drop feed back into collateral, covenants, or margin?*

If you can't sketch that loop in plain language, you don't yet know what you own. The Victorian era's harsh gift is clarity: great innovations can still produce terrible investments when they're financed in a way that cannot survive a single season of fear.

4.4 The Great Depression and Decades of Debt

The most frightening thing about a deflationary crash isn't the day the market breaks. It's the month after, when the headlines quiet down but your debts get heavier anyway. In a normal recession, prices might fall a bit and wages might wobble, but the number on your loan statement stays the same and you grit your teeth until demand returns. In a deflationary bust, the economy becomes a vice: incomes shrink, prices slide, and the *real* value of what you owe rises—like a treadmill that speeds up while you're already tired.

That's why 1929 America and late-1980s Japan belong together. The settings couldn't look more different—Jazz Age exuberance versus fluorescent-lit corporate Japan—but both ended up trapped in the same cruel geometry: too much debt secured by inflated assets, followed by falling prices that turned private balance sheets into engines of contraction.

When leverage stops being a tool and starts being a trap

Debt is not automatically dangerous. It's a practical invention: it lets you buy a house before you've saved for thirty years, build a factory without waiting a generation, or expand a business when opportunity arrives. The problem is leverage-debt used to amplify returns-because leverage assumes that tomorrow will be liquid, friendly, and willing to refinance you.

As long as asset prices rise, leverage feels like intelligence. You control more with less. Your equity-the slice you truly own-seems to grow quickly. The crowd starts confusing a favorable wind with sailing skill. Then prices stop rising, and leverage reveals its real personality: it is a contract with no sense of humor.

This is where the earlier panic loop returns with a sharper edge. A plain old crash is painful; a leveraged crash is destabilizing. When collateral values fall, lenders demand repayment or more collateral. Borrowers sell to meet those demands. Those sales push prices down further. The feedback loop is not merely psychological. It's contractual. Pieces of paper force actions.

Deflation adds a second blade. If prices broadly fall, the dollars you use to repay debt become harder to earn. Even if your interest rate is unchanged, your burden rises in real terms. That's the heart of the debt-deflation idea: indebtedness plus falling prices can create a self-reinforcing downward spiral, because the act of trying to become solvent (selling assets, cutting spending) can shrink the overall economy and push prices down further.

1929: the crash was the opening act

The popular memory of 1929 is cinematic: tickers rattling, fortunes evaporating, a great fall from the top of the chart. The more consequential story is slower and, in its own way, more horrifying: the financial system's plumbing failed in a way that shrank money and credit, and deflation did the rest.

The United States entered the 1930s with households and businesses that had borrowed during the boom years and a banking system that was not designed to absorb mass fear. When banks failed, deposits vanished for ordinary people. That wasn't a line item; it was cash for groceries, rent, payroll. Bank runs and failures didn't just reflect the Depression. They helped manufacture it by destroying credit and by intensifying uncertainty. When you don't know whether your bank will open tomorrow, you don't buy a refrigerator today.

Deflation is often described as "prices going down," which sounds almost pleasant—like a permanent sale. In a debt-heavy economy, it is closer to a silent wage cut combined with a surprise tax on borrowers. If you owe $1,000 and your income falls while the general price level falls, the *real* weight of that $1,000 rises. Multiply that effect across millions of borrowers and you get a society-wide scramble to reduce debt at the same time. The scramble looks responsible at the individual level-pay down what you owe, spend less, sell assets. In aggregate, it is disastrous. Everyone's attempt to become safer makes everyone else poorer.

And it wasn't only households. Businesses that might have survived a mild downturn found themselves crushed by collapsing revenues and tighter credit. Farms were squeezed by falling commodity prices. Manufacturers cut wages and jobs, which reduced demand further, which worsened business conditions. The economy became a hall of mirrors where each defensive move reflected back as a new problem.

A key amplifier in this era was the impairment of credit intermediation-plain English: the system that turns savings into loans stopped working properly. A functioning financial system is a quiet miracle. It evaluates borrowers, pools risks, and channels money to productive uses. When that system is damaged—by bank failures, by panic, by uncertainty—creditworthy borrowers can't get credit, and marginal borrowers get cut off abruptly. The difference matters because economies don't run only on "good ideas"; they run on working capital, inventories, payroll cycles, and ordinary financing.

This is why the Great Depression feels like a different species of event. It's not just that stocks fell or that output shrank. It's that the mechanisms that normally help an economy heal-credit creation, refinancing, confidence in payment systems-were impaired at the same time deflation was tightening the noose.

Japan in the 1980s: the richest bubble with the longest hangover

If 1929 is remembered as tragedy, Japan's late-1980s bubble is remembered as a kind of surreal extravagance. Tokyo land prices and the Nikkei's climb became shorthand for a world where money had lost its anchor. And for a while, Japan looked unstoppable: world-class manufacturing, disciplined firms, a rising standard of living, global admiration—and a financial system that, beneath the polish, was gorging on collateral values.

Japan's bubble had its own logic. The country had strong industrial capabilities and a high savings rate. Banks lent aggressively, often against real estate and equities. Companies and individuals borrowed because asset prices were rising, and asset prices were rising partly because borrowing was plentiful. It felt stable because it was widespread and because major institutions were involved. Stability, in bubbles, is frequently just unanimity.

Then the air leaked out. Equity prices fell sharply from their peak around the turn of 1990. Real estate prices followed with a lag, and that lag mattered: it kept the collateral illusion alive long enough for debts to remain large even as the underlying asset values deteriorated. A slow-moving collapse can be more damaging than a fast one because it keeps hope alive and delays decisive restructuring.

Japan's post-bubble era is often summarized with one phrase—"Lost Decades"—but the phrase can mislead. It wasn't a single, continuous catastrophe; it was a prolonged period in which the economy struggled to regain momentum because balance sheets were damaged and the financial system was cautious, sometimes paralyzed. When borrowers are over-leveraged, their priority shifts from expansion to repair. They divert cash flow to paying down debt rather than investing or hiring. When banks are nursing bad loans, they become conservative, even when interest rates are low. The result is a slow grind: not the dramatic collapse of a panic day, but a long season of subdued demand.

This is the essence of a balance sheet recession: when the private sector is focused on reducing debt, traditional stimulus—especially very low interest rates—can lose power. Even if borrowing is cheap, borrowers may not want more debt, and lenders may not want more risk. The economy becomes less responsive to the usual levers because the problem is not primarily the price of money; it's the desire to avoid obligation.

Japan's experience also complicates a comforting belief: that rich, sophisticated countries are immune to long stagnations. Japan had advanced industry, capable institutions, and high human capital. None of that guaranteed a quick recovery once debt and asset prices turned into a long-term drag. Prosperity reduces certain risks. It does not repeal financial gravity.

Why deflationary busts scar for so long

A useful way to think about these two episodes is that the initial crash is like a broken bone, and debt-deflation is like an infection. You can survive the break and still lose the limb if the infection spreads.

Three forces make these busts linger:

First, deflation redistributes wealth in a way that is economically toxic. It transfers purchasing power from borrowers (who tend to spend) to lenders (who may not). It turns default into a rational option for some and a moral crisis for others. It makes every contract more contentious because yesterday's terms no longer fit today's incomes.

Second, damaged financial intermediaries weaken the transmission of policy and the allocation of capital. Even if central banks cut rates or governments spend, the effect is blunted if banks are unwilling or unable to lend, or if borrowers are unwilling or unable to borrow. An economy can have abundant savings and still suffer from a shortage of *usable* credit.

Third, expectations change. People learn a new fear: not just that prices can fall, but that they can fall for a long time. That fear becomes self-fulfilling. If consumers and firms expect weak demand and declining prices, they delay purchases and investment. Delays reduce demand. Reduced demand validates the expectation. Confidence doesn't "return" in a single news cycle; it has to be rebuilt through repeated evidence.

This is why the most severe deflationary busts are not merely financial events. They are social events. They reshape attitudes toward risk, toward debt, toward institutions. They create political pressures that can last for a generation.

The trap of "it can't happen here"

Every era has its favorite lullaby: *We are more sophisticated now.* The song changes lyrics—better regulation, smarter central banks, improved risk models, deeper capital markets—but the melody is consistent. It's the same confidence that rose with canals, railways, and empires. The details can be true and the conclusion can still be wrong.

The deeper lesson from 1929 and Japan is not that history repeats perfectly. It's that leverage can create fragility in any society that allows debt to grow faster than incomes while collateral prices are rising. When that leverage is paired with a fall in the general price level—or even a long period of disinflation that raises real debt burdens—the adjustment can become slow and punishing.

Deflation is especially dangerous because it weaponizes time. In an inflationary world, debt burdens erode gradually, giving borrowers a tailwind toward solvency (though at the cost of purchasing power for lenders). In a deflationary world, the wind blows the other way. Waiting does not heal; waiting tightens.

What to watch in real markets: the quiet indicators of a long hangover

The signature of a debt-driven boom is not only exuberant prices. It's a change in how people justify those prices. You start hearing that refinancing is "easy," that rates will "stay low," that collateral is "as good as cash," that losses are "temporary," that defaults are "contained." These phrases are not proof of danger by themselves. They are flags that the system is leaning on continued smooth functioning.

In a deflationary bust, the clues shift. Watch for broad-based price weakness, falling nominal incomes, and a financial sector that is shrinking its balance sheet. Watch for a private sector that is hoarding cash not because it has no opportunities, but because it is repairing. Watch for policy that feels oddly ineffective: rates low, yet growth stubbornly weak. That combination is the smell of balance-sheet damage.

Most of all, watch for the moment when the conversation changes from "How high can it go?" to "Who's going to refinance?" That's the pivot from optimism to survival.

Actionable takeaway: run a personal debt-deflation drill

The most practical thing an investor can do with these historical scars is to rehearse them on a small scale—using their own finances as the model.

Once a year, do a "debt-deflation drill" in writing:

- Assume your household income (or business revenue) drops 10% and does not recover for two years.

- Assume the market value of your most debt-linked asset (home, rental property, leveraged portfolio, business equity) drops 20%.
- Assume refinancing becomes meaningfully harder: higher rates, stricter underwriting, or lower appraisals.

Then answer, concretely: *Which payments become stressful? Which assets would you be forced to sell? How long could you hold without selling at a bad price?*

If the exercise makes you uncomfortable, that's the point. Deflationary busts don't usually destroy people by surprise math; they destroy them by forcing sales when time is expensive. The antidote is not prediction. It's resilience: lower leverage than your optimism wants, more liquidity than your pride prefers, and a plan for what you will do if "temporary" turns into "two years."

Chapter 5

Modern Meltdowns and the Policy Response

Welcome to the era of the 'flash crash' and the trillion-dollar bailout. In this chapter, we leave behind the slow-moving disasters of the past to explore how technology and globalization have turbocharged financial panic. We will see how a computer glitch can erase billions in minutes, how a suburban housing dream turned into a global nightmare, and ask the critical question: when the government steps in to save the day, are they actually setting the stage for the next, bigger collapse?

5.1 The Speed of Greed: From Black Monday to the Dot-Com Bubble

On the morning of October 19, 1987, you could have walked onto the floor of the New York Stock Exchange and seen something that looked like competence: suits, badges, phones, paper tickets, a choreography refined over decades. Then the choreography broke. Prices began dropping fast enough that the human machinery of trading—people shouting, writing, running orders from one place to another—couldn't keep up. That mismatch between *how quickly investors could decide to sell* and *how slowly markets could absorb the selling* is the first big clue to modern meltdowns. The danger isn't just fear. It's fear with a fast trigger and a slow exit.

The popular myth of a crash is a single cause: some terrible piece of news, a villain, a bubble finally popping. Black Monday is unsettling because it doesn't grant that comfort. Nothing happened that weekend that could plausibly explain a one-day collapse of that magnitude. What changed was the *speed* and *mechanics* of the market. By the late 1980s, finance had discovered a new kind of leverage: not just borrowing money, but borrowing *liquidity*—the ability to assume you can sell when you want, at something close to the last traded price. That assumption works beautifully in calm weather. In a storm, it becomes a trapdoor.

A useful way to picture leverage is the classic margin loan: you put up $50, borrow $50, and buy $100 of stock. If the stock falls 10%, you lose 20% of your equity. Leverage doubles your emotional stakes. But the more lethal kind of leverage in a crash is *implicit* leverage: strategies that promise to limit losses or stabilize performance *as long as markets remain liquid*. The moment everyone tries to use the same "emergency exit," the exit narrows, and what looked like protection becomes gasoline.

That was the promise of *portfolio insurance*, one of the signature innovations of the era. The pitch was elegant enough to feel like physics: if the market starts falling, sell futures (or stocks) to reduce your exposure; if the market rises, buy them back. In theory, you're dynamically replicating a put option—a contract that pays off when prices fall—without having to buy the option itself. It sounded like a seatbelt that could be stitched on after you start driving. But unlike a seatbelt, portfolio insurance doesn't merely protect the wearer. It changes the traffic pattern for everyone on the road.

Here's the uncomfortable detail: portfolio insurance required selling into declines. Not because managers were panicking, but because the model told them to. The model was calm. The market was not. When enough large institutions followed the same rulebook, a down move became a signal to sell more, which made the down move bigger, which triggered more selling. Feedback loops are dangerous in any machine; in markets, they turn price changes into commands.

The crash also exposed a second, more mundane fragility: the plumbing. It's easy to imagine a market as a single scoreboard with a single price. In reality, 1987 was already a network: cash stocks on the NYSE, stock index futures in Chicago, options elsewhere, all linked by traders trying to exploit small differences between them. Those traders are called *arbitrageurs*, and in normal times they knit markets together by buying the cheaper instrument and selling the more expensive one. In a fast-moving panic, the knitters become yanked around by the yarn. Quotes go stale, orders pile up, confirmations lag, and the simple act of figuring out "the price" becomes a lagging indicator.

That matters because a lot of computerized and semi-computerized strategies rely on timely prices. If a futures price says the market is down sharply but the cash market hasn't updated yet (because it's overwhelmed), the mismatch can trigger program trades designed to close the gap. The intention is stabilizing: bring prices back into alignment. The effect can be destabilizing: pile more orders onto the clogged part of the system. A modern crash isn't always a stampede toward the door. Sometimes it's the door mechanism failing while the crowd keeps walking.

Black Monday is often described as the first crisis where machines led the selling. That's not quite right—there were humans everywhere, making choices, setting rules, approving models. The more accurate lesson is subtler and more unsettling: *humans moved the decision into the past*. They pre-committed. They wrote down what they would do *later* and outsourced the "later" to a procedure that didn't get tired, didn't negotiate, and didn't pause to ask whether everyone else was doing the same thing.

If you want a single phrase for the modern style of meltdown, it's this: *forced selling*. Forced selling is what happens when you don't sell because you've changed your mind about the future, but because you've lost the right to wait. Margin calls, risk limits, redemptions, collateral demands, model rules—these are financial gravity. They pull at the worst possible time, when prices are already falling and liquidity is already thin.

The decade after 1987 didn't remove that gravity; it refined it. By the late 1990s, Wall Street had become remarkably good at manufacturing leverage that didn't look like leverage. The packaging improved. The language softened. Borrowing became "financing." Concentrated bets became "relative value." The goal wasn't simply to be right; it was to be right *with more size.*

Long-Term Capital Management (LTCM) is the cautionary tale here, not because it was sloppy, but because it was sophisticated in exactly the way that tempts people into a false sense of security. The fund made trades that often looked modest: buying one bond and shorting a similar one, betting that the spread between them would narrow. That kind of trade feels almost humble compared to the stereotype of wild speculation. The danger was the scale. When the difference you're betting on is small, you lever the position to make the small difference matter. A basis point becomes a paycheck. A few basis points become a fortune. And a few basis points moving against you becomes an existential problem.

The seductive part of these "spread" trades is that they often look safe in historical data. The spread has narrowed before; it will probably narrow again. But probability is not protection when you're levered enough that "unlikely" is still too frequent. In a crisis, correlations jump. Assets that used to move independently begin to move together, not because their economics suddenly match, but because investors are selling what they can, not what they want to. In that environment, "diversification" can become a synonym for "many ways to be forced out at once."

LTCM's near-collapse in 1998 highlighted another modern accelerant: *funding fragility.* A highly levered portfolio doesn't just depend on the long-term value of the assets; it depends on the short-term willingness of other people to keep lending against them. When lenders get nervous, they ask for more collateral or they stop lending entirely. That creates a brutal rhythm: sell assets to raise cash, which pushes prices down, which makes lenders more nervous, which forces more selling. The story sounds like emotion, but it's largely arithmetic.

By the time the dot-com era arrived, the machinery of modern markets had matured: faster trading, more derivatives, more leverage, more institutions using similar risk systems, and a growing faith that the market itself was a kind of safety net. Then the internet arrived with a narrative powerful enough to make even prudent people feel slow. The bubble didn't begin with cynicism; it began with a truth: the internet really was transformative. That truth acted like a permission slip for bad habits. When a technology changes the world, it becomes easy to believe that *every* company associated with it is destined to win, and that the old ways of valuing businesses are obsolete.

The dot-com bubble is often told as a morality play about silly websites and sock puppets. That's entertaining, but it misses the more interesting mechanism: *how a good story changes what feels like risk.* When prices are rising and everyone around you seems to be getting richer, not owning the thing begins to feel like the riskiest position. The fear of missing out is not just envy; it's a reframing of danger. The bubble turns caution into regret.

What made dot-com special wasn't that people became irrational for the first time. It was that the market developed new ways to keep the party going even when skepticism was obvious. Consider short selling, the classic way to bet against an overvalued stock: you borrow shares, sell them, and hope to buy them back later at a lower price. In principle, a bubble invites short sellers the way honey invites flies. In practice, shorting hot internet stocks was often difficult, expensive, or outright impossible. Shares were hard to borrow. Borrow fees could be punishing. And the biggest practical problem was psychological: when a stock can double in a week, a short position can wipe you out long before "reason" returns.

That's the underappreciated point about bubbles: it's not enough for smart people to see the overvaluation. They have to be able to *express* that view in the market without getting carried out. Markets don't just aggregate beliefs; they also reflect who is constrained. If the optimists can buy with ease while the pessimists can't short effectively, prices can levitate longer than any spreadsheet says they should.

There was also a built-in supply mechanism that created its own drama: lockups. When a company goes public, insiders and early investors are often restricted from selling their shares for a period of months. During the mania, those lockups acted like a dam: a large block of potential selling pressure held back by contract. When the lockups expired, the dam cracked. Suddenly the market had to absorb a wave of shares from people who were already rich on paper and had every incentive to convert paper wealth into real money. In a market powered by optimism, new supply can be a pin to the balloon.

It's tempting to think the dot-com crash was simply the moment reality arrived. But reality was always there. The more precise description is that the market ran into its own internal limits: the ability to raise new money, the patience of buyers to absorb new issuance, and the growing gap between narratives and numbers. When that gap becomes embarrassing, the story loses its spell. And when the story loses its spell, leverage and forced selling do the rest of the work.

The link between 1987 and the late-1990s mania isn't "computers made people greedy." It's that computers and modern finance changed the *shape* of greed. In earlier eras, a mania might take years to inflate and years to unwind, because participation and execution were physically slow. By the end of the twentieth century, the cycle could compress. Money could move faster, and so could panic. The crucial transition wasn't just electronic trading. It was a world where many players outsourced judgment to systems that reacted to the same inputs in the same way.

That sameness is an underrated risk. Risk models, portfolio rules, and institutional constraints are often sold as safety features: discipline, governance, best practices. They are. But when everyone's discipline points in the same direction at the same moment, discipline becomes crowd behavior with better branding. A stop-loss rule is sensible for one investor. A stop-loss rule adopted by the whole market is a potential air pocket.

Another way to see the progression is through the idea of "liquidity as a product." In a calm market, liquidity feels like a natural resource—like air. You don't notice it until it's gone. Modern finance learned to treat liquidity as something you can rely on, monetize, and even synthesize through derivatives. Portfolio insurance implicitly assumed it could sell when it needed to. Levered spread trades implicitly assumed they could roll funding. Dot-com day traders implicitly assumed there would always be a bid. These are different stories with the same fragile hinge: someone else will be there to take the other side.

When that "someone else" disappears, the market stops behaving like a continuous auction and starts behaving like a series of gaps. That's when crashes feel uncanny: not a smooth decline, but sudden drops, broken correlations, and prices that look like mistakes. People often describe those moments as "no buyers." That's not quite true. Buyers exist; they are simply demanding a discount big enough to compensate for the risk of catching a falling knife. In other words, buyers are there, but at prices sellers can't emotionally accept until they're forced to.

The practical investor's lesson from these episodes isn't "avoid technology stocks" or "never use derivatives." The better lesson is to treat *speed* as a form of leverage. The faster you can trade, the easier it is to build positions you can't unwind under stress. The more automated your strategy, the more likely it is to behave in a perfectly logical way that becomes collectively disastrous. The more you depend on continuous liquidity, the more you should plan for discontinuous markets.

A final, concrete way to bring this home is to think about your own financial life as a balance between *optional* decisions and *forced* decisions. Optional decisions are the ones you get to make on your own timeline: rebalancing gradually, harvesting tax losses, adding to a position when it's down. Forced decisions are the ones the world makes for you: needing cash next month, holding an investment in a vehicle that can gate redemptions, borrowing against a portfolio that can trigger margin calls, owning something so illiquid that you can only sell by accepting a steep discount. Crashes turn optionality into a superpower.

Actionable takeaway: Make a personal "forced-selling audit." List anything in your financial life that could force you to sell at the wrong time—margin debt, short-term spending needs funded by volatile assets, concentrated positions you *must* liquidate on a schedule, or investments with limited liquidity. Then reduce just one of those pressures: extend your cash runway (even by a month), reduce leverage, or rebalance away from the position you'd be most tempted to defend emotionally. The goal isn't to predict the next crash; it's to make sure you won't become its involuntary seller.

5.2 House of Cards: The Global Financial Crisis

A mortgage is supposed to be the least cinematic form of finance. You buy a house. You pay every month. Life goes on. In the early 2000s, that boredom became a sales pitch: *home loans are safe because people will do anything to keep a roof over their heads*. And if the loan itself sounded too humble for Wall Street's tastes, no problem—it could be engineered into something that looked like a Swiss watch: diversified, rated, sliced into tidy layers, and sold around the world.

The global financial crisis didn't begin as a plot to blow up the economy. It began as a set of incentives that made ordinary decisions feel harmless. A mortgage broker who was paid per loan had a reason to approve one more borrower. A bank that could sell the loan had a reason not to worry too much about whether the borrower could pay in ten years. An investor chasing a little extra yield had a reason to believe a triple-A label meant "sleep well." Each step, by itself, was understandable. Together, they formed a system that could manufacture risk while hiding it from almost everyone who touched it.

The trick was to turn thousands of messy, human borrowers into a product that could be priced like a bond. That product was the *mortgage-backed security* (*MBS*): a bond-like instrument created by pooling many mortgages and passing the homeowners' monthly payments through to investors. The idea wasn't new. What changed was the scale, the loosened underwriting standards, and the confidence that home prices would keep rising. When you assume houses only go up, a delinquent borrower isn't a disaster—they're a speed bump. You refinance them, or you sell the house, or you roll the problem forward. The system starts to feel like it has no off switch.

An MBS pool is the first layer of abstraction. A borrower in Florida becomes a line item in a spreadsheet. Multiply that by thousands, and the borrower's story disappears. That disappearance can be useful: diversification is real, and pooling can reduce the impact of one unlucky job loss. But it can also be anesthetic. When the people who approve loans no longer expect to live with the consequences, the definition of "good enough" quietly slides.

Then came the feature that made the machine roar: *tranching*. An MBS could be split into slices—called tranches—with different priorities for getting paid. The top slice gets paid first and is designed to be very safe; the bottom slice absorbs the first losses and is designed to be risky. This is not magic; it's a waterfall. If homeowners keep paying, everyone gets their share. If homeowners stop paying, the lower tranches get hit first, protecting the upper tranches. The upper tranche can genuinely be safer than the underlying loans, *as long as losses remain within the buffer*.

That "as long as" was the crack in the foundation. The buffer is calibrated using models, and models need assumptions. The most important assumption was that housing markets in different regions wouldn't collapse together. A few neighborhoods? Sure. The entire country? Historically rare. Investors were, in effect, buying a bet that the American housing market would not experience a nationwide fever.

Once you can carve a pool of mortgages into slices with different ratings, you can sell "safe" yield to conservative investors and "spicy" yield to gamblers. And once there's a market for spicy yield, Wall Street can do something even more creative: bundle the spicy slices into a new product and tranche them again. That was the *collateralized debt obligation* (CDO) in its most notorious form. The marketing was irresistible: turn the lower-rated pieces of many MBS pools into a fresh structure that, through diversification and tranching, could again produce a large top slice rated triple-A. Risk, it seemed, could be recycled into safety.

Recycled is the right word, because the underlying substance didn't change. If you take the smoke detectors from ten houses that are on fire and build a new house out of them, you don't get a fireproof home. You get a very confident alarm system.

The other ingredient was *credit default swaps* (CDS), essentially insurance-like contracts that paid out if a bond defaulted. Used carefully, CDS can hedge real risk. Used aggressively, they allow investors to take on synthetic exposure—placing bets on the performance of bonds without owning them. That's where the crisis begins to resemble a crowded theater with a single exit: the number of claims can grow much larger than the underlying pile of mortgages. When losses arrive, the demand for protection and the obligation to pay it can overwhelm the institutions in the middle.

None of this would have detonated the way it did if the system had been funded with stable, boring deposits like a traditional bank. But much of it wasn't. It lived in the *shadow banking* system: bank-like activities (lending, maturity transformation, leverage) happening outside the perimeter of traditional bank regulation. Shadow banking wasn't necessarily illegal or hidden; it was "shadow" because it relied on wholesale funding markets—short-term borrowing from institutional lenders—rather than insured retail deposits.

This funding choice matters because of one simple, brutal mismatch: the assets were long-term and hard to sell (mortgages, MBS, CDOs), while the funding was short-term and flighty (overnight lending, short-term repurchase agreements, money market funding). That mismatch is called *maturity transformation*: borrowing short and lending long. It's a core function of banking. It's also why banks are fragile. When confidence is high, it looks like efficiency. When confidence breaks, it looks like a run.

A traditional bank run is vivid: depositors lining up. A shadow-bank run is quieter: lenders refusing to roll over short-term funding, demanding more collateral, or accepting only the safest securities. It happens on terminals and in conference calls. But the logic is identical. If you fund a long-term asset with short-term borrowing, you're promising you can always refinance. When lenders say no, you're forced to sell. Forced selling pushes prices down. Falling prices make lenders even more nervous. The loop tightens.

The years before 2008 were full of small signals that, in hindsight, were the creaks of a house taking on too much weight. Underwriting standards loosened. "No-doc" and "low-doc" loans spread—mortgages where income verification was minimal. Adjustable-rate mortgages with teaser rates made monthly payments look artificially affordable. Appraisals became optimistic. Meanwhile, home prices rising year after year created a dangerous emotional proof: *the system is working*. When a strategy produces money, it becomes easy to confuse luck and design.

The real fuel was leverage. Not just homeowners taking on big mortgages, but institutions using borrowed money to hold these securities. Leverage is seductive because it turns small spreads into big profits. If you can borrow cheaply and earn a little more by holding a "safe" triple-A structured product, you've built a money machine. The only catch is that the machine depends on the product staying "safe," the funding staying cheap, and the ability to sell staying continuous. All three can fail at once.

The ratings were the social glue that held the structure together. A pension fund in Europe might not be allowed to buy anything below a certain rating. A money market fund might need top-quality collateral. A bank's regulatory capital requirements could be influenced by ratings. So the rating wasn't merely an opinion; it was a key that unlocked entire pools of buyers. When that key is handed out too freely, the market's immune system gets tricked into welcoming something it should question.

When housing prices began to flatten and then fall, the story changed from "a few borrowers will have trouble" to "the collateral itself is deteriorating." Delinquencies rose. Defaults rose. The lower tranches started absorbing losses—which they were designed to do. The problem was that the losses didn't stop at the lower tranches, because the assumptions about correlation were wrong. Many regions weakened together. Refinancing became harder. The exit door narrowed.

As confidence cracked, the crisis moved from credit losses to liquidity panic. That shift is important. A system can survive losses if it can fund itself while it works through them. What it can't survive is a sudden refusal by short-term lenders to keep lending. In 2007 and 2008, institutions discovered that "liquid" securities can become illiquid when everyone is trying to sell them, and that models can't force a buyer to appear.

There's a human way to understand this without a single equation. Imagine you own a house, and you've been told your neighborhood is stable and your home value will hold. You take out a home equity loan because it seems prudent. Then a few houses on your street sell for much less than expected. Suddenly, the bank wants to re-evaluate your collateral. You haven't missed a payment. You haven't done anything wrong. Yet you're being squeezed because the market's perception of your asset has changed. Multiply that sensation across an entire financial system, and you get the emotional texture of 2008: institutions being punished not only for what they *did*, but for what they *held*.

The securitization chain is where the crisis becomes easiest to picture, because it shows how a single mortgage payment can be transformed into an international promise.

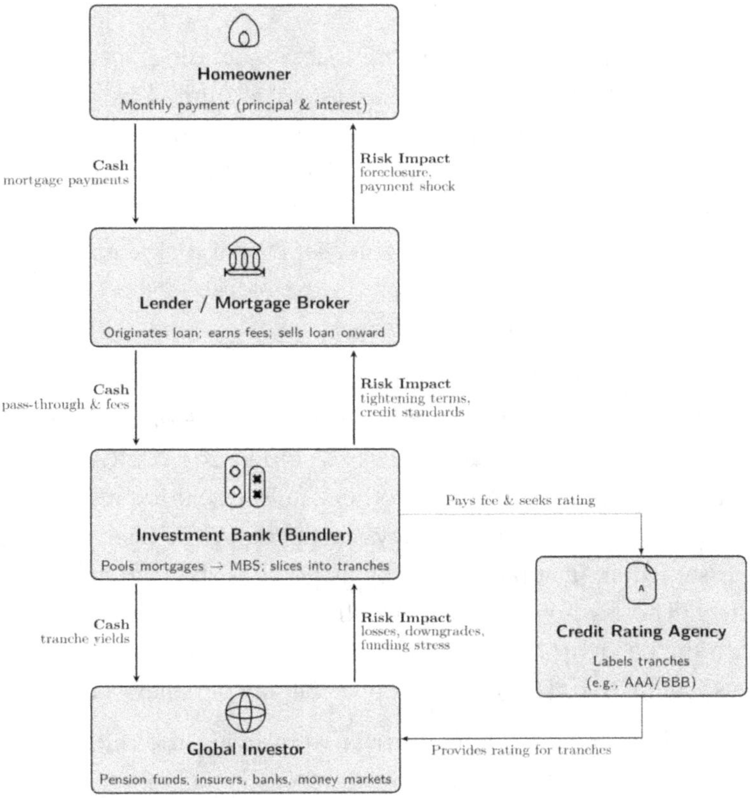

The diagram looks clean. The real-world version was anything but. Each arrow had contracts, legal entities, off-balance-sheet vehicles, servicing agreements, and hedges layered on top. Complexity wasn't just a byproduct; it was a feature. Complex structures create plausible deniability. When something goes wrong, it becomes harder to answer the most uncomfortable question in finance: *who exactly owns the loss?*

That question turned poisonous in 2008 because many institutions couldn't convincingly answer it about themselves. If you own a plain bond, you can mark it to a market price and move on. If you own a bespoke structured product and the market for it evaporates, pricing becomes a debate. When pricing becomes a debate, trust becomes scarce. And modern finance runs on trust the way an engine runs on oil.

This is where "shadow banking" stops sounding academic and starts sounding personal. If the system is funded by short-term lenders who can leave, then survival depends on confidence. Confidence depends on transparency. Transparency was exactly what the system had traded away for yield. When investors realized they didn't understand what they owned—or what their counterparties owned—they behaved rationally in the oldest way possible: they tried to get their money back first.

The crisis accelerated through a series of institutional scares that taught the market a new reflex: *don't wait for clarity*. Bear Stearns' collapse sent the message that even a storied firm could lose access to funding quickly. Lehman Brothers' failure sent the message that the government might not always catch you. AIG's distress sent the message that "insurance" is only as good as the insurer's balance sheet. These weren't merely headlines; they were lessons written into the behavior of lenders. After each shock, the market became less willing to roll funding, accept collateral, or believe ratings.

Meanwhile, ordinary homeowners were living the micro version of the same dynamic. Teaser rates reset. Home values fell below mortgage balances. Refinancing options vanished. Foreclosures rose, pushing neighborhood prices down, creating more negative equity, creating more defaults. It's the same feedback loop again, just with lawns and moving trucks instead of term sheets.

The phrase "toxic assets" became popular because it captured a social reality: certain securities didn't just lose value, they became embarrassing to hold. In finance, embarrassment is contagious. If you fear that holding a thing will make your lenders doubt you, you sell it. If everyone sells it, the price falls. If the price falls, holding it becomes even more embarrassing. Liquidity vanishes not because the asset ceases to exist, but because nobody wants to be seen owning it.

One of the most misunderstood features of 2008 is that the losses alone were not the whole catastrophe. Losses were the spark; *funding* was the oxygen. A highly levered system can absorb only so much mark-to-market pain before it triggers collateral calls and deleveraging. Once deleveraging starts, it can become self-fulfilling: selling pushes prices down, which forces more selling, which pushes prices down again. Even sound assets get dragged lower because they're the easiest to sell.

If 1987 taught that machines can accelerate selling, 2008 taught that *balance sheets* can accelerate it too. When leverage meets short-term funding, speed becomes structural. You don't need a computerized strategy to be forced out—you just need lenders who can demand cash today against assets that can only be sold at a discount tomorrow.

The crisis also revealed how moral language sneaks into financial debates. People asked, with understandable anger, why borrowers took loans they couldn't afford, why bankers made deals that looked predatory, why investors believed ratings that seemed too generous. Those questions matter. But focusing only on individual blame misses the system's most important feature: it made bad behavior feel normal and good behavior feel uncompetitive. When a firm that tightens standards loses market share to firms that don't, restraint begins to look like incompetence.

A practical way to understand this is to imagine being a cautious lender in 2005. You require full documentation, meaningful down payments, and conservative appraisals. Your competitor across town offers low-doc loans with teaser rates and quickly sells them into securitizations. Borrowers flock to the competitor. Your volumes drop. Your shareholders ask why you're "missing the boom." That's how bubbles recruit participants: not by hypnotizing them, but by punishing those who refuse to play.

For everyday investors, the crisis offers a deeper lesson than "housing can fall." It teaches that the risk in a financial product is not just the risk of the underlying asset, but the risk of the *funding model* wrapped around it. An MBS is tied to homeowners, yes. It's also tied to the willingness of institutions to lend against it, the rules that govern who can hold it, and the market infrastructure that lets it be priced. In a panic, the wrapper can fail before the contents.

Actionable takeaway: When you evaluate any investment that promises bond-like stability with stock-like returns, ask one unglamorous question before you ask about yield: *who is providing the funding, and what happens if they leave?* If the answer involves short-term borrowing, leveraged vehicles, or "we can always sell," treat the investment as riskier than its label suggests. In your own finances, preserve optionality the same way a cautious bank does: keep a real cash buffer for near-term needs, avoid borrowing against volatile assets, and be skeptical of products whose safety depends on continuous liquidity.

5.3 The Domino Effect: Contagion Beyond Borders

If money had a passport, it would be a forgery. The modern financial system moves capital across borders with the casual ease of an email, but it moves fear even faster. A bank can be headquartered in Paris, funded in dollars borrowed overnight in New York, hedged with derivatives cleared in London, and stuffed with bonds issued in Rome. When a shock hits one part of that chain, the question isn't "who caused it?" so much as "who needs cash *today*?" In crises, geography becomes trivia and funding becomes destiny.

Contagion is the polite term. It sounds like a metaphor, as if markets catch colds. The reality is more mechanical: balance sheets are connected by contracts, and contracts have clauses that wake up at the worst possible time. Margin requirements rise, collateral gets rejected, credit lines get pulled, and a problem that looked local suddenly creates a global demand for the same thing: safe collateral and reliable cash. The crowd doesn't stampede because it's irrational; it stampedes because the math forces everyone toward the same narrow door.

A useful way to think about cross-border crises is to separate two kinds of "spreading." The first is a *common shock*: oil prices spike, global growth slows, interest rates rise, a pandemic shuts down travel. Many countries suffer at once, but not because they infected each other. The second is true *transmission*: one country's problem forces another country to act, sell, or default. This is the domino effect: not simultaneous pain, but sequential toppling.

The 1997 Asian Financial Crisis is one of the cleanest demonstrations of transmission because it begins with something that looks safe and ends with something that looks like chaos: a currency peg. A currency peg is a promise that your money will trade at a fixed rate to another money—often the U.S. dollar. For businesses, it feels like stability. For governments, it can feel like a badge of seriousness. For foreign lenders, it's a green light: if the exchange rate won't move, lending in dollars to a Thai company seems almost as safe as lending in dollars to an American one.

That apparent safety created a classic mismatch. Many companies and banks across the region borrowed in dollars (because the interest rates were lower and the money was plentiful) while earning revenue in local currency (baht, rupiah, won). This is called *currency mismatch*, and it is financial horror hiding in a spreadsheet. As long as the peg holds, the mismatch is invisible. If the local currency falls, the borrower's debt effectively balloons overnight. Imagine your mortgage doubling because the unit of account changed. That's the kind of stress that turns a solvency problem into a political crisis.

Why did the peg fail? Because a peg is not merely a promise; it's a promise backed by reserves and credibility. If investors believe the central bank has enough foreign currency reserves—and enough will—to defend the peg, they won't test it. If they suspect the opposite, testing the peg becomes profitable. Speculators can borrow the local currency, sell it for dollars, and wait. The central bank buys its own currency to keep the price fixed, draining reserves. The act of defense becomes evidence of weakness. When reserves thin out, the peg snaps. "Contagion" begins to look less like a disease and more like a bank run conducted in foreign exchange.

Once Thailand devalued, the region learned the lesson it didn't want: pegs can break. Investors who had treated "emerging Asia" as a single asset class suddenly began selling across countries, even those with stronger fundamentals. Some of that selling was lazy, the way humans are lazy under stress: it's faster to hit the "risk-off" button than to do careful country-by-country analysis while prices are moving. But much of it was forced by institutional plumbing. Hedge funds facing redemptions sold what they could. Banks reduced exposures to preserve capital. Currency traders raised margins. Risk managers tightened limits. The crisis spread because the system's rules compressed diverse economies into a single risk bucket at the exact moment nuance was most valuable.

There's a cruel irony here. When global money is flowing in, it treats developing countries as unique stories: the "Asian miracle," the "tiger economy," the "new middle class." When the tide turns, it treats them as tickers. "Emerging markets" becomes one word, and that one word gets sold.

Contagion also travels through trade, not just finance. A currency collapse makes a country's exports cheaper and its imports more expensive. Neighbors suddenly face tougher competition and weaker demand. Firms that sold components into regional supply chains see orders vanish. Governments respond with higher interest rates to defend their currencies, which slows domestic growth, which hurts tax revenues, which weakens confidence further. The feedback loop has a local flavor, but the mechanism is the same as in any panic: actions taken to protect stability in the short term can deepen fragility in the medium term.

Fast-forward a decade, and the global system faces a different kind of border-crossing fragility: not pegs, but a shared currency. The Euro was designed to remove a classic source of European crises—competitive devaluations and exchange-rate volatility. It succeeded in that. It also removed a pressure valve. If a country shares a currency, it cannot devalue to regain competitiveness, and it cannot create currency to backstop its banks in the same way it could with its own central bank. The Eurozone Sovereign Debt Crisis wasn't simply about irresponsible governments; it was about a monetary union that created tight financial integration without fully integrating fiscal policy and banking backstops.

Before the crisis, the Euro made different countries' government bonds feel almost interchangeable. Greek bonds didn't pay *quite* the same yield as German bonds, but the spread was small enough to look like a mild personality difference rather than a fundamental risk difference. Banks across Europe loaded up on sovereign debt.

Some did it for yield. Some did it because sovereign bonds are often treated as high-quality collateral. Some did it because regulators and risk models encouraged the idea that government debt, in its own currency, is special.

Then came the realization that broke the spell: euro-area governments issue debt in a currency they do not individually control. Greece could not print euros to pay bondholders. Italy could not unilaterally devalue. Suddenly "sovereign risk" returned to a continent that had been pricing it like an antique.

That's when the Eurozone's signature contagion machine switched on: the *doom loop* between banks and governments. Here's how it works in plain language. Banks hold a lot of their own government's bonds. If investors begin to doubt the government's ability to repay, bond prices fall. Falling bond prices hurt banks' balance sheets, because the assets they hold are worth less. A weaker banking system makes the economy weaker—credit tightens, investment slows—which hurts the government's tax revenues and increases its need to spend. The government looks weaker, bond prices fall further, banks get hit again. The loop feeds itself.

Governments then face a choice that feels like a trap. If they rescue banks, they take on more debt and look less creditworthy. If they don't rescue banks, the banking system can implode, crushing the economy and making the government look less creditworthy anyway. Investors watch this dilemma unfold and demand higher yields to compensate for the uncertainty. Higher yields make the debt burden heavier. The "price" of borrowing becomes part of the crisis, not a reflection of it.

This is where contagion becomes political. In a country with its own central bank, there is at least a theoretical backstop: the central bank can act as lender of last resort, stabilizing funding markets and preventing self-fulfilling runs. In the Eurozone, that backstop was complicated by governance and legitimacy. Who should bear the cost?

Who should set the conditions? What is solidarity, and what is moral hazard? These are not questions markets enjoy waiting to resolve.

The Eurozone crisis also revealed another cross-border reality: banks can be national in their regulation and politics, yet transnational in their exposures. A German bank can hold Greek bonds. A French bank can have loans to Spanish developers. An Irish bank can be funded through markets that turn skittish because of events in Italy. When stress hits, each national government is tempted to protect its own institutions, but those institutions are linked to everyone else. It's like trying to fight a fire with doors open between apartments.

A separate but equally powerful channel of modern contagion runs through the world's dominant funding currency: the U.S. dollar. This surprises many people, because it sounds like a geopolitical trivia fact. It's not. A vast amount of global borrowing and lending is denominated in dollars, even when neither party is American. A European bank may take deposits in euros but make loans or hold securities in dollars. To fund those dollar assets, it needs dollar funding—often short-term. In calm times, dollar funding is plentiful. In a panic, everyone wants dollars at once.

When non-U.S. institutions scramble for dollars, they often sell assets to raise cash or try to borrow in stressed funding markets. That selling pressure can hit U.S. markets, emerging markets, and anything liquid enough to be turned into dollars quickly. This is how a crisis that begins in one corner of the world can show up as "weird" price moves in another. It's not always about beliefs. It's about settlement and collateral.

Central bank swap lines became one of the most important, least-understood crisis tools for this reason. A swap line is essentially a standing arrangement where the Federal Reserve can provide dollars to another central bank (say, the European Central Bank), which then lends those dollars to its own banks. It's an emergency bridge across a currency mismatch. Without it, banks can be forced into fire sales simply because they can't obtain the currency they need to meet obligations. Swap lines are contagion medicine: they don't cure the underlying disease, but they can stop a liquidity panic from turning into mass insolvency.

There's a pattern in all these episodes that's worth noticing because it changes how you interpret headlines. Crises spread fastest through *standardization*. The more the world uses the same funding currency, the same collateral rules, the same risk models, and the same market conventions, the more quickly it can synchronize into a single global trade: sell risky things, hoard safe things. Integration makes markets efficient in good times. In bad times, it makes them move as one organism.

It also changes what "diversification" means. Many investors believe diversification is a geographic trick: own some U.S. assets, some European assets, some emerging markets, and you're protected. Geography helps when shocks are local. But contagion events are explicitly designed to ignore borders. In a global panic, correlations rise not because economies suddenly become identical, but because investors are forced to raise cash in the same way, from the same types of holdings, at the same time. You can hold ten countries and still be exposed to one global margin call.

This doesn't mean international investing is futile. It means the risk you need to manage isn't merely "country risk." It's *funding risk*. Who is financing the positions? In what currency? With what maturities? Against what collateral? These questions sound institutional, but they matter to individuals because your mutual fund, your bank, your pension plan, and your employer all sit inside that same web.

If you want a personal, concrete way to feel contagion, consider the experience of an export business during a currency crisis. Your customers abroad suddenly become poorer in your currency. Orders slow. Your local bank, worried about its own funding, tightens credit. Your costs don't fall as fast as your revenue. You didn't trade derivatives or buy foreign bonds, yet you're living inside a cross-border balance sheet event. Contagion is often experienced by ordinary people as "the air changed," as if opportunity simply thinned. Underneath, it's the tightening of credit and the retreat of risk-taking.

The most actionable insight is also the least exciting, which is exactly why it's useful. When crises go global, the winners aren't the people with the best hot takes about which country is "next." The winners are the people who have optionality: cash reserves, stable funding, and a portfolio that doesn't require selling at the bottom to meet near-term needs. Contagion punishes anyone who needs the market to cooperate on a particular Tuesday.

Actionable takeaway: Stress-test your life for a borderless shock. Assume that in a severe global event, stocks fall together, credit tightens, and the U.S. dollar strengthens (making foreign returns feel worse in dollar terms, at least temporarily). Then make one adjustment that increases your resilience without requiring prediction: extend your cash runway for essential expenses, reduce any borrowing tied to volatile assets, and avoid concentrating your emergency savings in a bank or instrument that could be indirectly exposed to wholesale funding stress. The goal is not to outguess contagion; it's to be the person who doesn't have to sell into it.

5.4 The Safety Net Paradox: Bailouts and Moral Hazard

A bailout is finance's most emotionally charged word because it collides with two instincts that rarely coexist peacefully: the desire to punish recklessness and the desire to stop a fire from spreading. In the middle of a panic, those instincts don't take turns. They wrestle. One voice says, "Let them fail." The other says, "If they fail, what else breaks?" The safety net paradox lives in that second question: the more believable the rescue, the less careful the system may be *before* the emergency.

The modern economy runs on promises that are only occasionally visible. Your payroll hits your account because banks trust other banks overnight. A company can make payroll because it can roll short-term funding. A pension fund can rebalance because markets stay open and prices remain continuous. When those promises wobble, the problem is not just losses. It's coordination failure: everyone wonders whether everyone else will still perform. At that point, "bailout" is often a sloppy label for something more technical and more basic: restoring belief that tomorrow's payments will happen.

Central banks were invented for this kind of moment. The classic role is *lender of last resort*: when private lenders won't lend, the central bank lends to solvent institutions against good collateral, at a rate that discourages casual use. It's a deliberately unglamorous job description. The goal isn't to make money or pick winners; it's to stop a liquidity panic—a shortage of cash and trustworthy funding—from turning into a solvency disaster where otherwise-viable firms collapse because they can't refinance on a Tuesday.

That distinction between *liquidity* and *solvency* is where many public arguments go off the rails. Liquidity means you can't get cash quickly enough. Solvency means your assets, in a realistic sense, are worth less than your liabilities. In a calm world, the difference is clear. In a panic, it becomes philosophical, because asset prices gap down, markets freeze, and "realistic" becomes a moving target. A bank can look insolvent at fire-sale prices and solvent at hold-to-maturity prices. Policymakers have to decide in real time which set of prices should be treated as truth.

That's why crises produce emergency programs that look, from the outside, like improvised plumbing. The Federal Reserve and other central banks create lending facilities with specific rules: eligible collateral, haircuts (a discount applied to protect the lender), interest rates, and term lengths. The names change, but the logic doesn't. If a key market freezes—commercial paper, repo funding, asset-backed securities—the central bank can step in as a temporary buyer or lender to restart circulation. The intent is not to subsidize bad assets forever; it's to prevent forced selling and cascading defaults from destroying the financial infrastructure that healthy businesses rely on.

A useful mental picture is a crowded stadium. Most days, people exit calmly. In a panic, even a perfectly safe stadium can become deadly if everyone rushes the same gate. A lender-of-last-resort program is not a gift to the people who built the stadium badly; it's a way to open additional gates before the crowd crushes itself. The uncomfortable part is that once stadium-goers learn the gates will open, some builders may become less careful about designing exits in the first place.

Fiscal bailouts add a different kind of power, and a different kind of controversy. When governments use taxpayer-backed resources to inject capital into banks, guarantee liabilities, or purchase troubled assets, they're doing more than lending. They're absorbing risk that private markets refuse to bear. That can be stabilizing. It can also feel like an injustice with a balance sheet.

The Troubled Asset Relief Program (TARP) is the signature American example of this tension. In the fall of 2008, the financial system wasn't merely bruised; it was losing its ability to fund itself. TARP was authorized to provide a large fiscal backstop, and in practice it became a tool for recapitalizing banks and stabilizing key institutions and markets. The political optics were brutal: rescuing banks while unemployment rose and foreclosures spread. Yet from a systemic perspective, recapitalizing banks wasn't charity; it was triage. Under-capitalized banks can't lend, and an economy without lending is an economy on crutches.

This is where the safety net paradox stops being abstract. If you rescue the system aggressively, you may prevent a depression. You may also teach investors and executives that the downside of correlated risk-taking will be shared. In finance, shared downside changes behavior. If people believe the state will step in to prevent the worst outcomes, they may rationally choose strategies that are individually profitable and collectively dangerous.

That collective danger often takes the form of *correlated leverage*. Imagine two strategies. Strategy A is boring: modest leverage, long-term funding, assets that can be sold without wrecking the market. Strategy B is exciting: high leverage, short-term funding, assets that are safe in normal times, and a plan to refinance continuously. Strategy B pays more in tranquil years. If a crash is severe enough, Strategy B blows up—unless the state provides emergency liquidity or guarantees. If the state reliably provides that backstop, Strategy B becomes even more attractive. Multiply that logic across many institutions, and the system becomes full of Strategy B, all leaning on the same fragile funding structure. The safety net that prevents catastrophe today can, if expected, increase the probability of catastrophe tomorrow.

Moral hazard is the label for this behavior shift. The clean definition is simple: *when someone is protected from consequences, they may take more risk.* Insurance companies know this. Seatbelt laws know this. Financial policy knows this too, but the financial version is harder because the "insured" party is often not a single person. It's an entire ecosystem of banks, funds, and investors adjusting their risk choices based on what they expect policymakers will tolerate.

In practice, moral hazard doesn't require executives twirling mustaches. It can be subtle and spreadsheet-shaped. Risk managers model "stress scenarios" based on historical events and plausible policy responses. Traders price assets assuming some probability of rescue in extreme cases. Creditors lend at lower spreads because they believe the state won't allow a disorderly failure. Those lower spreads make leverage cheaper. Cheaper leverage makes balance sheets bigger. Bigger balance sheets make failures more dangerous. More dangerous failures make rescues more likely. The loop is tight, and it is rational at every step.

The policy problem, then, is not whether to have a safety net. Modern finance without a safety net tends to discover, the hard way, that runs and cascades are part of its nature. The real problem is designing the net so it catches falling bodies without turning high-wire acts into a sport for amateurs.

How do you do that? First, by drawing a bright line between protecting the *system* and protecting the *owners*. This is why many crisis interventions are designed to lend against collateral rather than give grants, and why bank rescues often include dilution of shareholders and constraints on dividends, bonuses, or buybacks. The principle is straightforward: keep payments flowing, keep credit arteries open, but don't preserve equity value as if it were sacred. If capital is wiped out when a firm fails, investors will demand discipline in advance.

Second, by making emergency support expensive enough that nobody uses it for convenience. Central bankers sometimes call this lending at a penalty rate. The idea is to keep the facility as a backstop, not a subsidy. In calm times, private markets should beat the central bank on price. In panic times, the central bank should be available on terms that prevent a death spiral.

Third, by forcing institutions to pre-build their own shock absorbers: more capital (a bigger equity cushion), more liquid assets, less reliance on runnable short-term funding, and clearer resolution plans. Capital is the simplest form of resilience because it doesn't require anyone else's permission. A well-capitalized firm can absorb losses without triggering immediate funding stress. Liquidity buffers allow a firm to survive a period when refinancing becomes difficult. Resolution plans—sometimes called "living wills"—are attempts to ensure that a complex firm can be unwound without taking the payment system with it.

The most ambitious attempt to address the bailout dilemma is to make failure *possible* again without making it *catastrophic*. That's the promise behind post-crisis resolution regimes that allow authorities to take over a failing institution, impose losses on shareholders and certain creditors, and keep critical operations running. When done well, this is neither a pure bailout nor a pure liquidation. It's controlled demolition: preserve the infrastructure, redistribute the losses to those who knowingly took risk, and prevent the panic from becoming a stampede.

The problem is that credibility is hard to manufacture. In quiet years, policymakers can insist that no firm is too big to fail. In a real panic, they face a different audience: creditors who can pull funding in minutes, investors who can dump assets at scale, and voters who will punish either action or inaction. If markets doubt that authorities will actually impose losses on powerful creditors, then funding remains cheap and moral hazard remains alive. If markets doubt that authorities will actually prevent disorderly collapse, then funding can vanish at the first hint of trouble. Policy has to walk a narrow ridge between complacency and terror.

It's also worth admitting something that rarely gets said plainly: bailouts can be unfair *and* necessary. Fairness and necessity are different questions. A rescue that stabilizes the economy may still offend basic moral intuitions, especially if the people who profited during the boom keep their gains while ordinary workers pay with layoffs and lost homes. That anger is not a misunderstanding; it's a valid diagnosis of how asymmetric modern finance can be. If policymakers ignore it, they pay later in political backlash and institutional distrust—which can, ironically, make future crises harder to manage because legitimacy becomes scarce.

The deepest version of the safety net paradox is psychological. A successful rescue often looks like an overreaction in hindsight. If the crisis didn't end in total collapse, skeptics conclude the collapse was never going to happen. This is the curse of prevention: the better you do, the less credit you get. But in markets, that skepticism feeds risk-taking. People remember the rescue, not the abyss it was meant to avoid. Over time, the memory of the abyss fades, while the expectation of the rescue hardens into a belief.

Investors can see this in miniature in their own behavior. After a long bull market with a few quick interventions—a cut in rates, a liquidity facility, a bailout—it becomes tempting to believe that downturns are temporary and that buying the dip is not a strategy but a law of nature. That belief can work for years. It can also set the stage for the moment when policy is slower, weaker, or constrained by inflation, politics, or credibility, and the market discovers that a safety net is not the same thing as a trampoline.

None of this leads to an easy moral conclusion, because the trade-off is real. If authorities refuse to intervene, the result can be a chain of failures that harms people who never signed up for the risk: employees, small businesses, retirees. If authorities intervene too freely, they can create a financial culture that treats leverage as a one-way bet. The goal is not purity. The goal is a system where private actors have strong reasons to be prudent, and public actors have tools to stop panics from becoming depressions.

For a practical reader, the most valuable takeaway isn't how to judge policymakers. It's how to judge your own exposure to the policy cycle. When the safety net expands, risk assets often rise and credit spreads compress. When the safety net is questioned, everything that relied on easy funding can reprice violently. If your financial plan requires permanent support—low volatility, continuous liquidity, painless refinancing—you're implicitly betting on policy, even if you never buy a bond.

Actionable takeaway: Build a "no-rescue" version of your plan. Assume that in a severe downturn, policymakers respond slowly, political fights delay action, and markets stay stressed longer than expected. Then adjust one element of your finances so you can endure that scenario: reduce any personal leverage (especially margin or variable-rate debt), increase your liquid emergency buffer, and avoid concentrating your wealth in instruments that depend on daily liquidity while holding hard-to-sell assets underneath. If the safety net catches the system, you'll still be fine. If it doesn't, you won't be forced to make your worst decision on the worst day.

Chapter 6

Navigating the Future: New Frontiers and Lasting Lessons

We often convince ourselves that we are smarter than our ancestors, armed with sophisticated algorithms, instant data, and global connectivity. But while our financial tools have evolved, the human brain has not. In this concluding chapter, we look at how the ancient demons of fear and greed manifest in the age of hashtags and high-frequency trading. We will explore how social media accelerates mania, why the 2020 crash rewrote the rulebook on government intervention, and ultimately, how you can build a personal survival kit for the inevitable storms ahead.

6.1 The Viral Markets: Meme Stocks and Crypto Cycles

A century ago, joining a mania required geography. You had to be near the rumor, near the broker, near the crowd. Today the crowd lives in your pocket, and it never sleeps. A notification vibrates: a stock is "moving." Another: a coin is "trending." A third: your friend just posted a screenshot of a one-day gain that looks like a lottery ticket. The modern mania doesn't ask you to travel to the frenzy; it delivers the frenzy to you, frictionless, personalized, and—crucially—*interactive.*

That interactivity changes the emotional physics of speculation. Classic bubbles were fueled by stories, but the stories moved at the speed of letters, newspapers, and dinner parties. Viral markets run on stories that can be copied, remixed, and blasted to millions in minutes. When attention can be scaled instantly, *excitement becomes a form of leverage*: not borrowed money, but borrowed momentum. The asset doesn't just rise; it rises *in public*, and the public rise becomes part of the product.

The GameStop episode of January 2021 made the mechanism visible in a way textbooks never could. A struggling brick-and-mortar retailer became the stage for a mass experiment in narrative, coordination, and market plumbing. The plot had everything: a villain (short sellers), a hero (the "little guy"), a rallying cry ("hold the line"), and a scoreboard that updated every second. When the price surged, it didn't merely reward early buyers—it recruited new ones. The chart itself became marketing, and every spike was an advertisement that someone else was getting rich *right now*.

To understand why that matters, it helps to translate a few pieces of market jargon into plain English. A *short sale* is a bet that a stock will fall: you borrow shares, sell them, and hope to buy them back cheaper later. It's not inherently sinister; it's how many investors express skepticism, and it can expose fraud. But it has a built-in vulnerability: if the price rises instead of falls, the short seller faces potentially unlimited losses and may be forced to buy shares to close the position—buying that can push the price up even more. When short interest is unusually high, the market contains dry tinder. All it needs is a spark, and in 2021 the spark was social media's ability to turn a trade into a cause.

What made the moment uniquely modern wasn't only the story of "Wall Street versus Main Street." It was the way the story *constrained choices*. When a trade becomes identity—when "buy" becomes "belong"—selling stops feeling like a financial decision and starts feeling like betrayal. That's not a moral critique; it's a description of how crowds behave when money and meaning fuse. In older manias, you could quietly change your mind. In viral manias, the crowd watches you do it. Your brokerage app makes it one tap to buy, and your online community makes it psychologically expensive to sell.

Then there was the accelerant that turned a bonfire into a firestorm: options. An *option* is a contract whose value depends on an underlying asset. A call option, for instance, gives you the right (not the obligation) to buy a stock at a set price. Options can be used responsibly, like insurance. But in a viral market they can behave like rocket fuel because they allow a small amount of money to control a large amount of exposure. That's leverage wearing a party hat.

Options also create mechanical feedback loops through hedging. Many option dealers hedge their risk by buying the underlying stock as the stock price rises, especially when lots of call options are bought. This hedging can amplify upward moves: rising price leads to more hedging demand, which can push the price higher, which triggers more hedging. It's not a conspiracy; it's plumbing. But when the plumbing meets a stampede, pressure builds fast. Viral markets are where psychology and microstructure—human emotion and mechanical hedging—can lock arms and sprint together.

That interplay is why meme-stock manias can feel like they have a soundtrack. The beat is the refresh rate. Price moves create posts; posts create attention; attention creates trades; trades create price moves. In the analog era, the loop was slowed by the time it took to place orders, transfer cash, read tomorrow's paper, or call your broker. In the app era, the loop is tight enough to hum. The market doesn't merely *reflect* sentiment; it *manufactures* sentiment in real time, and then sells it back to you as "what everyone is watching."

There's a darker twist: the same loop that can lift a stock can also pull it apart. Viral attention is fickle. The internet doesn't "invest"; it binge-watches. The crowd that can coordinate buying pressure can also coordinate boredom. And boredom is lethal to a narrative-priced asset, because narrative is the scaffolding that holds up valuations when fundamentals are not doing the heavy lifting. When the story loses oxygen, what remains is the price—and the price, without story, looks suddenly indefensible.

Crypto cycles took that viral dynamic and made it a permanent feature rather than a rare episode. Crypto markets run 24/7, which means the emotional thermostat has no night. There is no closing bell to force people to stop staring at the chart, no weekend to let adrenaline drain, no pause where professionals can reset risk limits and retail investors can go outside and remember they have hobbies. When a market never sleeps, neither does the urge to check it—and the urge to check it is a form of participation. Even the decision *not* to trade is made under the pressure of constant, visible motion.

The core promise of crypto has often been technological—new rails for money, programmable contracts, censorship resistance. But the lived experience for many participants has been social and gamified. Tokens come with mascots, memes, slogans, and communities that feel less like shareholder bases and more like sports fandoms. "Winners" are celebrated publicly; "paper hands" are mocked; skepticism is treated as negativity rather than analysis. The market becomes a stage where your financial decisions are also your social signals.

Leverage in crypto has its own peculiar glamour. In traditional markets, leverage is often hidden behind margin agreements, institutional funding desks, and risk committees. In crypto, leverage can be wrapped in a clean interface and offered as a default setting, sometimes with cartoonish badges and celebratory animations. The psychological effect is subtle but powerful: risk starts to feel like a game mechanic, not a real-world amplifier of loss.

And because crypto markets are global and lightly bounded by time zones, the crowd is always large enough to become contagious. A rumor doesn't need a trading floor; it needs an algorithm. A tweet can become a macro event. A single exchange's outage can become a systemwide panic. In that environment, "liquidity"—the ability to buy or sell without moving the price too much—can evaporate fast. When everyone tries to exit through the same digital door, the door doesn't widen just because the building is virtual.

The most instructive crypto collapses are the ones that mimic old-fashioned bank runs, because they reveal that technology can change the wrapper without changing the human substrate. A *stablecoin* is designed to hold a stable value—often one U.S. dollar—so it can function as cash inside crypto markets. Some stablecoins are backed by reserves, like short-term government securities or cash. Others have attempted stability through engineering: mechanisms that try to maintain the peg through incentives and market operations.

The engineering approach can work—until confidence breaks. Stability is, in a sense, a social agreement. Once enough people believe a peg might fail, they race to redeem first. The race itself becomes the reason the peg fails. That is the oldest story in finance: a run doesn't require insolvency at the start; it can *create* insolvency by forcing rapid liquidation or by overwhelming the mechanism meant to provide safety. Crypto did not invent this dynamic; it rediscovered it at high speed.

The Terra/Luna collapse in May 2022 is a case study in reflexive fragility: the stablecoin's promise depended on a linked token's market value, and that market value depended on confidence in the stablecoin. Reflexivity is a fancy word for a simple idea: in some systems, the output changes the input. When prices rise, confidence rises, which can push prices higher. When prices fall, confidence falls, which can push prices lower. In reflexive systems, the feedback is not a bug; it's the engine. The danger is that engines can run in reverse.

Crypto also illustrated an institutional lesson that feels uncomfortably familiar: when markets are booming, people mistake *access* for *safety*. A sleek interface, a well-known sponsor, a stadium naming deal, a celebrity endorsement—these are signals of social proof, not guarantees of solvency. Social proof is useful for choosing restaurants. It is a dangerous substitute for due diligence in finance because it points you toward what is popular, not what is robust. In manias, popularity and fragility often travel together.

Viral markets didn't just speed up speculation; they changed who gets the first draft of reality. In older eras, prices moved and journalists explained afterward. Now commentary and price discovery are braided together. The explanation arrives before the move ("watch this"), during the move ("we're doing it"), and after the move ("here's why it had to happen"). When narrative is continuous, it can become self-sealing: every outcome is interpreted as validation. If the price rises, the thesis is "proven." If the price falls, the thesis is "attacked," and the attack becomes a reason to double down. That's how communities turn volatility into ideology.

This is where democratization gets complicated. Zero-commission trading and easy onboarding are, in principle, wonderful. Lower fees and broader access can help people build wealth. But "access" can also mean *access to mistakes*—faster, bigger, and with fewer speed bumps. When an app makes trading feel like texting, the boundary between entertainment and risk dissolves. The market becomes another feed, and the feed is optimized for engagement, not for your long-term financial health.

Engagement thrives on immediacy. Investing thrives on patience. Those are opposing design goals. A well-designed entertainment product trains you to seek novelty, check frequently, and react quickly. A well-designed investment process trains you to tolerate boredom, check infrequently, and react slowly. Viral markets are what happen when entertainment design principles are applied to financial decisions that punish impulsivity.

It's tempting to blame the crowd—the "dumb money," the "kids on the internet," the "degenerates." That temptation is comforting because it implies that experience is immunity. Yet the history of finance is not a story of foolish amateurs repeatedly tricking wise professionals. It's a story of humans—amateurs and professionals alike—becoming overconfident together. Professionals can be swept up by incentives: career risk, performance pressure, the fear of underperforming peers, the desire to look innovative. Retail investors can be swept up by belonging, hope, and the very rational wish to escape financial anxiety. Viral markets simply synchronize these motives more efficiently.

A useful way to see the difference between a "normal" market and a viral one is to focus on what sets the price. In a fundamentals-driven market, new information about cash flows, competitive dynamics, or economic conditions does most of the work. In a viral market, *flows*—who is buying and selling, how urgently, and with what leverage—can dominate in the short run. That doesn't mean fundamentals don't matter; it means the timeline shifts. Fundamentals are patient. Flows are impatient. Viral markets privilege the impatient.

The practical question is not whether viral markets will recur. They will, because the technology that enables them is not going away, and because the human desires they harness—status, belonging, excitement, hope—are permanent. The practical question is how to participate in markets *without donating your nervous system* to the crowd.

One way to stay grounded is to keep a simple mental split-screen: *story* versus *structure*. The story is what people say: the memes, the mission, the visionary claims, the villain narratives. The structure is what makes the market behave: leverage, liquidity, concentration, settlement constraints, options positioning, the fragility of pegs, the quality of reserves. Stories are not useless—stories can point to real changes—but structure tells you how quickly a situation can turn from "exciting" to "untradeable."

Another grounding tool is to treat volatility as a bill that eventually comes due. Viral markets often feel like free energy: price movement you can harvest if you're clever enough, fast enough, early enough. But volatility is not just opportunity; it is also the mechanism that transfers wealth from the late and leveraged to the early and disciplined. When an asset's main feature is that it moves, the product you are buying may be *emotion* rather than expected return.

The sharpest edge of viral markets is not that people can lose money. People have always lost money in markets. The sharpest edge is that losses can be socially engineered to feel like heroism. When losing becomes proof of loyalty, risk management becomes shameful. That is how a market stops being a market and starts being a tribe. Tribes are excellent at creating meaning. They are terrible at setting stop-losses.

A concrete way to immunize yourself—at least partially—is to decide in advance what kind of participant you are in any high-velocity trade. Are you investing (buying because you believe long-term value will grow)? Are you trading (buying because you expect short-term price movement)? Or are you spectating (watching because it's thrilling)? All three are allowed. The trap is pretending you are doing one while behaving like another. Viral markets are full of people "investing" with a two-day time horizon and "trading" with no exit plan.

A small table can help crystallize the distinction, because clarity is often the first form of protection:

	Fundamentals-First Market	*Viral Market*
Dominant fuel	Earnings, cash flows, economic data	Attention, positioning, leverage, narrative
Typical tempo	Days to years	Minutes to weeks
Key risk	Being wrong about value	Being right about value but wrong about timing and liquidity
Common tell	Boring headlines, slow price drift	Explosive charts, constant commentary, identity language
Failure mode	Gradual disappointment	Sudden air pockets, forced selling, runs

The point isn't to sneer at viral markets. They can reveal genuine frustrations (like perceived unfairness), and they can accelerate adoption of real technologies. They can even create life-changing wins for some participants. The point is to recognize their distinctive hazard: when the feedback loop is tight, the crowd can turn on a dime, and the same social machinery that made buying feel easy will make selling feel impossible—right when selling is most urgent.

Actionable takeaway: before you enter any trade that is being driven by online excitement, write down (in one sentence) *why you are buying, what would prove you wrong,* and *the exact price or condition under which you will reduce or exit.* Then place your position size so that if the crowd moves against you overnight—because in viral markets it can—you can follow your plan without negotiating with your identity.

6.2 Systemic Shocks in a Globalized World

In early March 2020, the world learned a strange new phrase for an old human impulse: the *dash for cash*. It wasn't poetic, but it was accurate. The pandemic didn't just threaten lungs and hospital capacity;

it threatened payrolls, rent checks, corporate credit lines, municipal budgets—everyday obligations that don't care about epidemiology. When fear arrives with a calendar, people don't merely want safety. They want *spendable* safety. So they sold what they could, even if "what they could" was supposed to be the safest thing on earth.

That's the first uncomfortable lesson of systemic shocks: in a real scramble, the market doesn't politely separate "risky assets" from "safe assets." It separates "assets I can sell right now" from "assets that might trap me." In March 2020, stocks fell hard, of course. But the more revealing story was the stress that showed up in the plumbing—those quiet pipes beneath the headlines that move cash, collateral, and confidence through the financial system.

Modern finance looks sleek on the surface. Tap an app, buy a share. Swipe a card, pay for groceries. Behind the scenes, it's a global chain of promises: banks, funds, dealers, clearinghouses, money-market funds, repo markets, prime brokers, custodians, payment rails. When a shock hits, the chain tightens. Everyone asks the same two questions at once: *Do I have enough liquidity?* and *Will the next link in the chain still perform?* The first question drives selling. The second question drives runs.

The pandemic shock also arrived at an awkward moment. Corporate balance sheets were not uniformly fragile, but they were interconnected in ways that rewarded efficiency over redundancy. Nonbank finance—think hedge funds, bond funds, mortgage REITs, private credit vehicles—had grown larger and more important. Many of these players aren't "banks" in the legal sense, yet they can behave like banks in a crisis: they borrow short, invest longer, and rely on markets staying liquid. When markets stop cooperating, they don't have a branch network of insured deposits to stabilize them. They have to sell.

Selling under stress has a particular smell. It's not the deliberate decision-making of a long-term investor; it's triage. You sell the assets with the tightest bid-ask spreads. You sell what you can finance easily. You sell what your risk system tells you is suddenly too large. You sell because a margin call is a deadline. Margin is borrowed money against collateral; when collateral prices fall, lenders demand more collateral or less borrowing. Margin calls don't ask if you *believe* in your trade. They ask if you can wire cash by 2 p.m.

That's how a health crisis turned into a market microstructure crisis. The cascade ran through margin, leverage, and the forced logic of risk limits. It also ran through something the public rarely thinks about: dealer balance sheets. Dealers—big banks and broker-dealers—help absorb flows and make markets. But their capacity isn't infinite. Regulations, internal risk limits, and the simple reality of capital constraints can make dealers reluctant to take on inventory when everyone else is trying to offload it. When dealers step back, liquidity evaporates faster than most investors imagine possible.

The U.S. Treasury market is the best place to see this, precisely because it's the last place you'd expect trouble. Treasuries are the benchmark "risk-free" asset in global finance, used as collateral in countless transactions. They are supposed to be the place money hides when the world feels unsafe. Yet in March 2020, Treasuries experienced dysfunction: sharp, unusual volatility and signs that the market's ability to smoothly intermediate huge selling pressure was impaired. The paradox—*the safe asset acting unsafe*—was a clue that the problem wasn't about credit risk. It was about the mechanics of liquidity.

Why would anyone sell Treasuries in a panic? The answer is unromantic: because Treasuries are *what you can sell*. If you need dollars immediately—because clients are redeeming fund shares, because lenders want more margin, because you're meeting payroll, because you're staring at a collapsing revenue line—then the most liquid thing becomes your ATM. In a dash for cash, "flight to quality" can happen at the same time as "sell the quality to get cash." Those two impulses are not contradictory; they are just happening in different parts of the system.

Here's where globalization stops being a slogan and becomes a force multiplier. The dollar is not merely America's currency; it is a key funding currency for the world. Companies and institutions outside the United States often borrow in dollars, trade in dollars, and hold dollar assets. When global stress spikes, the demand for dollars can surge abroad. If offshore institutions can't easily obtain dollar funding, one way to raise dollars is to sell dollar-denominated assets—often Treasuries. That selling pressure can boomerang back into the U.S. market that everyone assumes will always function.

This is the moment where the "market" begins to look less like a set of prices and more like an ecosystem. You can't understand the equity selloff without understanding credit. You can't understand credit without understanding collateral. You can't understand collateral without understanding Treasuries and repo. You can't understand Treasuries without understanding global dollar demand. A systemic shock is not a single domino; it's a crowd pressing on the same exit, and discovering that the exit is narrower than it looked during the calm.

Central banks exist, in part, because we've learned the hard way that liquidity is a public good during panics. In March 2020, the Federal Reserve moved with a speed and scale that would have seemed unthinkable in earlier eras. The goal wasn't to prop up stock prices as an abstract act of kindness. The goal was to restore the functioning of markets so that forced selling didn't turn into a self-reinforcing liquidation spiral.

Think of liquidity support as a firebreak. In a wildfire, you don't negotiate with the flames. You remove fuel, create space, and stop the spread. In markets, the "fuel" is forced selling funded by short-term borrowing and amplified by margin calls and redemptions. The "space" is the ability of the system to absorb sales without collapsing prices. By injecting liquidity—through large-scale purchases, repo operations, and new lending facilities—the Fed aimed to widen the exit, reduce the need for fire sales, and stabilize the collateral that underpins so much borrowing.

The result, in the most visible market, was the now-famous *V*. The speed of the decline was breathtaking. The speed of the recovery was, in historical context, almost surreal. And that's where the second uncomfortable lesson lives: modern markets can be fragile *and* resilient in the same month. Fragile because leverage and interconnected plumbing can seize quickly. Resilient because a credible backstop can change behavior instantly. When investors believe a buyer of last resort will stand behind market functioning, the panic can reverse before the news gets less scary.

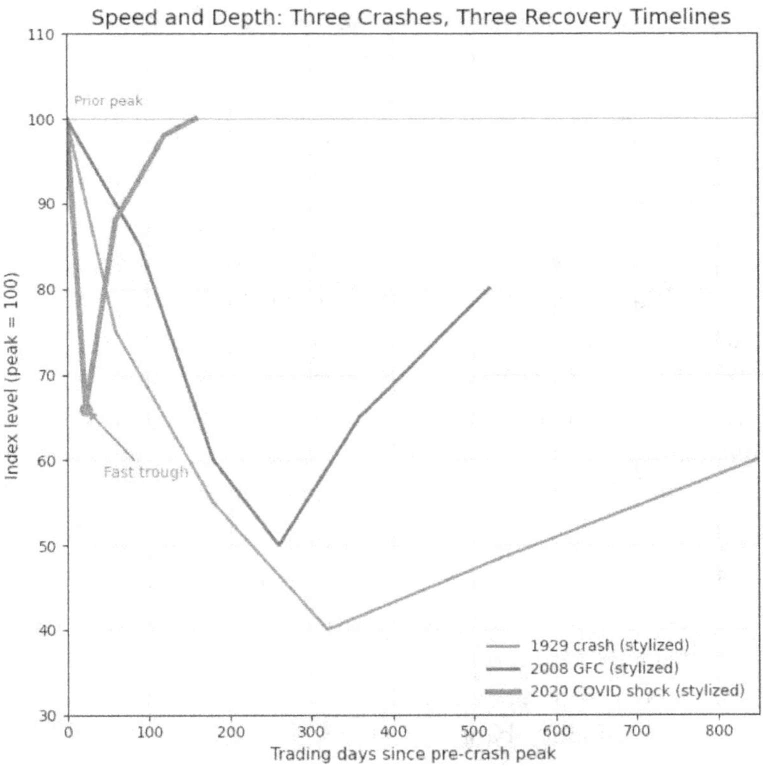

That chart is a reminder that time is part of the risk. A 35% drawdown that heals quickly feels very different from a 50% drawdown that lingers for years, even if the lowest point is "just a number" on a page. Humans don't experience losses in percentages; they experience them in sleepless nights, postponed retirements, and the quiet dread of checking an account balance again. The psychological scar tissue of a slow recovery changes behavior. The strange feature of 2020 is that the scar tissue formed in the crash—and then got abruptly confused by the rebound.

The rebound also taught a lesson that is both comforting and dangerous: the *belief* in support can become a market force of its own. Investors talk about the "Fed put," borrowing the language of options. A put option is insurance against downside; it pays off when prices fall. The metaphor suggests that when markets tumble, the central bank will step in to limit the damage. Whether that's a fair description depends on the moment and the mandate, but the *expectation* alone can change risk-taking. If investors internalize a safety net, they may climb higher on the ladder—until the day the safety net is smaller than they assumed.

That's moral hazard in plain clothes. It doesn't require greed or villainy. It can emerge from reasonable inference: "They stepped in last time, so they'll step in next time." The problem is that policy makers can't promise to support every asset price, and even when they act, there are lags, constraints, and political limits. A system that is stabilized by intervention can also become a system that quietly *needs* intervention more often, because participants build strategies that are viable only in a world where liquidity is abundant and volatility is contained.

March 2020 also sharpened the distinction between *solvency* and *liquidity*. Solvency means you have more assets than liabilities; you're fundamentally able to pay. Liquidity means you can meet obligations *on time* without selling assets at fire-sale prices. A solvent household can still go bankrupt if it can't pay next month's mortgage.

A solvent institution can still fail if it can't roll funding tomorrow morning. The panic of 2020 was, in many places, a liquidity crisis first—an urgent need for cash under conditions where selling anything large moved the price against you.

This matters because liquidity crises are contagious. When one institution sells aggressively, it pushes prices down, which triggers margin calls and risk-limit breaches for others, which forces them to sell, which pushes prices down further. The system becomes a feedback loop. This is the same basic rhythm as a classic bank run, but distributed across markets and mediated by collateral values rather than lines outside a branch.

The global dimension adds another layer of fragility: currency mismatch and cross-border funding. If you earn revenue in one currency but owe debt in dollars, you become sensitive to dollar funding stress. If your institution depends on short-term funding that can disappear, you become sensitive to confidence. In a crisis, confidence doesn't vanish evenly; it vanishes in pockets, and the pockets can be geographically far away from the original shock. A virus doesn't respect borders, and neither does a funding squeeze.

The Federal Reserve's international tools—swap lines with foreign central banks and the FIMA repo facility—were not esoteric diplomatic gestures. They were, in effect, attempts to prevent a global dollar shortage from forcing overseas holders to dump Treasuries into an already stressed market. This is the counterintuitive reality of a globalized system: helping liquidity abroad can be a form of self-defense at home, because the same collateral and funding markets tie everyone together.

Technology both helped and hurt. Electronic markets can process immense volume, but they can also transmit stress instantly. Risk systems can cut exposures in milliseconds, which sounds prudent until everyone's system reaches the same conclusion at once. The efficiency of modern finance can behave like synchronized swimming in calm water—and like synchronized panic when the pool catches fire.

If there's a single image worth keeping from March 2020, it's not a red stock chart. It's the idea of Treasuries—supposedly the world's smoothest market—getting clogged. That's the market telling you, in plain language, that "liquidity" is not a property of an asset. It is a property of an *ecosystem*: balance sheets, funding markets, regulations, risk appetites, and the willingness of intermediaries to stand in the middle.

That ecosystem can be strengthened, but it can't be made invincible. Regulators can push more activity into central clearing, improve transparency, and monitor leverage outside the banking system. Funds can stress-test for redemption waves. Dealers can hold more capital. But the basic truth remains: the modern system is built to be efficient most days, and efficiency is often the opposite of spare capacity. Spare capacity is what you want when the world goes sideways.

The practical investor doesn't need to memorize every facility the Fed created in 2020, or every acronym in the crisis alphabet soup. What's worth memorizing is a mindset: in systemic shocks, prices don't merely respond to "bad news." They respond to *financing constraints*. If you want to understand how a crisis will evolve, ask who is forced to sell, what they are forced to sell, and who has the balance sheet to buy it without becoming the next forced seller.

Actionable takeaway: treat liquidity as a personal risk factor, not a background assumption. Before the next shock arrives, decide what portion of your portfolio must be convertible to cash within a week without relying on a heroic market (high-quality cash equivalents, short-duration instruments, truly diversified sources), and size your riskier positions so you never *need* to sell them during a stampede. In a dash for cash, the best asset isn't the one with the highest expected return—it's the one that keeps you from becoming a forced seller.

6.3 Diagnostics: Spotting the Next Bubble

The most expensive sentence in investing is usually said with a grin: "It can't go down." Sometimes it's dressed up as sophistication—"structural support," "permanent demand," "a new regime." Sometimes it's plain bravado. Either way, it's the moment the mind quietly replaces probability with destiny. Bubbles don't announce themselves with confetti. They announce themselves with certainty.

The trick is that you don't need to predict the exact day the music stops to protect yourself. You need to notice when the room is getting crowded, when the exits are narrowing, and when the band is being paid in IOUs. A good diagnostic isn't a crystal ball; it's a smoke alarm. It won't tell you which ember becomes the fire. It tells you the air has changed.

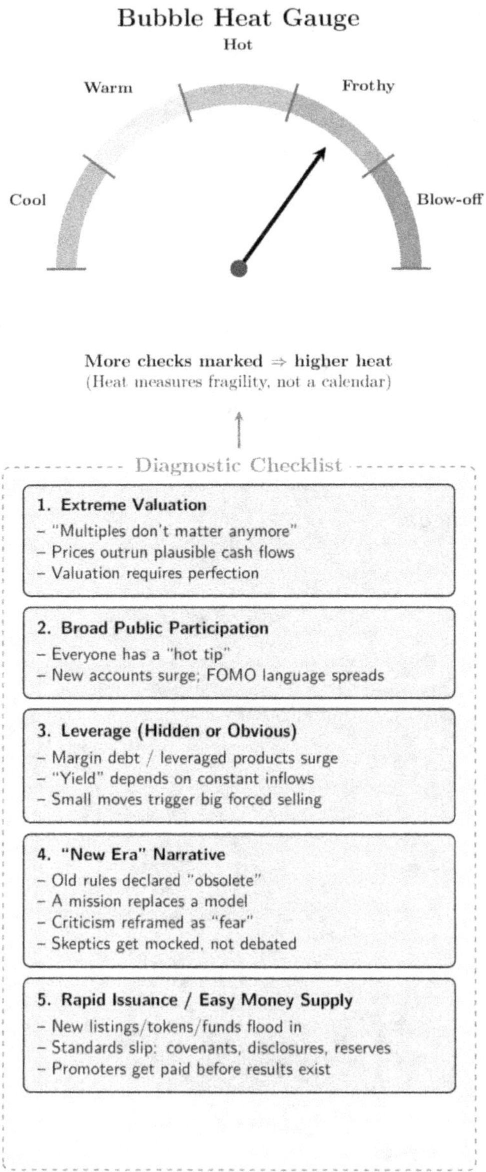

A diagnostic like this is deliberately unfair to your ego. It doesn't care how brilliant the story sounds, how charismatic the founder is, how elegant the technology looks, or how many smart people are buying. It asks a colder question: *Is the market becoming fragile?* Fragility is what turns a normal downturn into a cascade—when small bad news triggers forced selling, runs, or margin spirals.

The five signals above are not "reasons to short." They are reasons to stop pretending you're in a calm, fundamentals-driven world. A market can keep levitating long after it becomes fragile. If you treat "fragile" as synonymous with "imminent crash," you'll learn the wrong lesson and you'll hate diagnostics forever. Fragility is a condition, not a schedule.

1. Extreme valuation: the arithmetic starts to require miracles

Valuation is just a way of asking how much future success is already baked into today's price. In a bubble, the price doesn't merely assume things go well—it assumes things go well in a way that leaves no room for ordinary human failure: competition, regulation, execution mistakes, recessions, boredom. Perfection isn't impossible. It's just rare, and bubbles tend to price it as if it's the default.

An easy tell is when the conversation begins to treat valuation like a quaint superstition. You'll hear phrases that sound liberating: "It's about total addressable market," "It's about community," "It's about the network," "It's about the mission." Sometimes those phrases describe real value. The diagnostic question isn't whether they're *ever* true; it's whether they're being used as a spell to make arithmetic disappear.

If you want one practical move here, it's this: translate the price back into a plain-English requirement. "For this to make sense, this company must earn $X per year within Y years," or "this token must capture Z% of global payments." When you say it out loud, you often hear the absurdity—or, occasionally, you hear something surprisingly plausible. Either way, you've reclaimed the ground from vibes.

2. Broad public participation: the trade becomes dinner-table conversation

A bubble is not just expensive; it's social. At a certain temperature, investing stops being a private activity and becomes a public sport. People don't merely own the asset; they *perform* the ownership. Screenshots travel. Group chats light up. Neighbors become macro strategists. Barber shops, yoga studios, family reunions—everywhere becomes a trading desk.

This is not a sneer at newcomers. Broad participation is healthy in many contexts; capital markets should not be a gated community. The problem is that bubbles recruit at the moment when risk is most underpriced. When you see a wave of new entrants arriving specifically because recent returns look effortless, you're looking at demand that is sensitive to mood. Mood demand vanishes when charts turn red.

A clean diagnostic question: *Are people buying because they understand the asset, or because they understand that other people are buying it?* That second reason can work for a while. It's also the reason exits jam.

3. Leverage: the invisible hand becomes a clenched fist

Leverage is borrowed exposure. It makes gains feel genius and losses feel like gravity. In overheating markets, leverage tends to expand quietly because it's rewarded. Lenders loosen standards, margin becomes cheaper, and products appear that offer "enhanced" returns. Even when investors aren't explicitly borrowing, leverage can hide inside structures—derivatives, yield promises, short-term funding against long-term assets.

The diagnostic isn't "does leverage exist?" Leverage is part of modern finance. The diagnostic is whether leverage is becoming *casual*. When people treat borrowing as a lifestyle choice rather than a hazard, you're in a danger zone. When the pitch for an investment emphasizes how easy it is to lever up—or how "safe" the yield is—you should hear an alarm.

One practical way to feel leverage, rather than merely talk about it, is to imagine a forced-sale scenario. Ask: *If this drops 30% quickly, who is forced to sell?* Not who *should* sell. Who *must*. Leveraged players must. Funds facing redemptions must. Anyone pledged to daily liquidity while owning illiquid assets must. This is where systemic shocks are born: not from opinions, but from deadlines.

4. The "new era" narrative: when disagreement is treated as a character flaw

Every bubble has a slogan that sounds like emancipation. It's the promise that the old constraints—business cycles, valuation, cash flow, risk—have been defeated by innovation or insight. Sometimes the innovation is real. The bubble is the part where the innovation is used to justify any price.

Watch how the community handles criticism. In a healthy market, skepticism is annoying but informative; it forces better arguments. In an overheated market, skepticism is treated as malicious. Critics are not "wrong"; they are "haters," "dinosaurs," "captured," "anti-progress." That shift matters because it signals the market is losing its ability to self-correct. When a group can't tolerate dissent, it can't process new information—so it processes volatility instead.

A diagnostic question that cuts through the noise: *Is the narrative falsifiable?* Can anything happen that would make believers change their mind? If the answer is no—if every outcome is spun as proof—then you're no longer looking at analysis. You're looking at faith.

5. Rapid issuance: when supply appears because pricing is generous

When an asset class becomes a money machine, the machine attracts builders. Some builders are real entrepreneurs. Others are promoters. In a bubble, the market's generosity becomes a production incentive: more IPOs, more SPACs, more token launches, more funds, more "opportunities." New supply is not inherently bad—growing industries need capital. But the pattern is revealing: issuance tends to surge when buyers are least picky.

This is the "cash register" signal. If a sector can raise money easily, it will raise money *excessively*. Standards slip because standards slow the party. Disclosures get glossy. "Community" substitutes for governance. Reserves become ambiguous. Covenants weaken. The details that protect investors are framed as obstacles to innovation.

A quick, practical check: look at who gets paid and when. In healthy capital formation, insiders make most of their money when the business performs over time. In bubble issuance, insiders can get paid for *selling the dream*—through fees, tokens allocated at favorable terms, sponsor promotes, or early liquidity windows—long before the dream meets reality.

A bonus diagnostic: the credit mirror

Equity bubbles get the headlines, but credit often tells the more honest story. Credit is where optimism has to sign paperwork. When credit is cheap and abundant, it lubricates booms. When credit tightens, it exposes which business models were swimming naked.

A particularly useful macro lens is the idea of "excessive credit growth"—when borrowing grows much faster than the economy that is supposed to support it. The Bank for International Settlements popularized the *credit-to-GDP gap* as an early-warning indicator: roughly speaking, how far credit has drifted above its long-term trend relative to the size of the economy. You don't need to compute it yourself to use the intuition. If an asset boom is being financed by a rapid expansion of borrowing—especially outside traditional banks—you are watching fragility build.

This also helps explain why some bubbles feel unstoppable: credit is the oxygen. When oxygen is plentiful, fires spread. When oxygen is cut off, even the most beautiful narrative has to confront cash flow and refinancing dates.

How to use the gauge without turning into a doomed market timer

The hardest part of bubble diagnostics is emotional. Seeing froth makes people do one of two unhelpful things. They either become smug ("I'm smarter than the crowd") or they become hypnotized ("I need to get in before it's too late"). Both reactions are about ego. Neither is about risk management.

A better approach is to link the heat gauge to *positioning* rather than prophecy. If heat is low, you can afford to be patient and valuation-driven. If heat is rising, you can shrink position size, reduce leverage to zero, raise liquidity, and insist on a larger "margin of safety" (the cushion between price and what you think something is worth). If heat is extreme, you can shift from "how much can I make?" to "how do I make sure I'm still standing if this snaps back violently?"

This is also the moment to remember that "diversification" is not just owning many tickers. In bubbles, correlations rise—assets that seemed unrelated begin to move together because they are owned by the same levered players or because they are sold to meet the same margin calls. True diversification includes liquidity, duration, and exposure to different economic outcomes, not just different brand names.

A practical, non-glamorous tool from the professional world is the *pre-mortem*. Imagine that the trade ends badly. Not mildly badly—badly enough that you feel embarrassed telling someone about it. Then ask: what likely caused that outcome? Was it leverage? A liquidity freeze? A regulatory change? A stablecoin de-peg? A credit event? A cyber incident? A wave of redemptions? The point is not to be paranoid. The point is to identify *how* you could lose, because losses usually arrive through channels you did not bother to picture.

Finally, be honest about the role of entertainment. Some people participate in frothy markets the way they participate in sports betting: small stakes, clear limits, no illusions. That can be psychologically healthier than "investing" with your rent money while telling yourself it's a long-term thesis. The diagnostic gauge is not a morality test. It's a tool for setting stakes that match reality.

Actionable takeaway: pick one market you're exposed to—or tempted by—and score it using the five warning signs on the Bubble Heat Gauge. Then make a single concrete change that reduces fragility in your own finances *today*: cut any personal or portfolio leverage to zero, raise your cash buffer, or halve the position size of the most narrative-driven holding. If you can't name a change you're willing to make, you're not doing diagnostics—you're doing astrology.

6.4 The Cycle of Forgetting: Institutional Memory and Survival

Every crash eventually turns into a punchline. Not because the losses were funny-tell that to someone who postponed retirement or lost a home-but because time does what time always does: it sands down the sharp edges. The stories get shortened. The villains get simplified. The dread gets edited out. A decade later, the same market that once felt like a burning building becomes a place where people complain about how "nothing ever happens." Calm, in finance, is not neutral. Calm is a solvent that dissolves fear.

That's the cycle of forgetting in one sentence: when risk doesn't bite for a while, people stop believing it has teeth. They don't say it like that. They say "volatility has structurally declined," or "the system is better capitalized," or "this is a new era." Sometimes those statements contain real progress. What makes them dangerous is what they allow next: more leverage, thinner margins of safety, bolder promises, tighter funding, and a casual attitude toward liquidity—the exact ingredients that turn ordinary surprises into panics.

Generational turnover is the simplest engine behind the pattern. The people who remember the last inferno eventually leave the room. Traders retire. Risk managers get promoted into roles where their caution is treated as pessimism. Regulators rotate. CEOs who lived through a near-death experience are replaced by CEOs whose formative memory is a long bull market and a PowerPoint slide titled "resilience." Meanwhile, new entrants arrive with fresh confidence and no scar tissue. They are not stupid; they are *uninjured*. And uninjured people are often braver than they should be.

What's tricky is that forgetting doesn't look like amnesia. It looks like sophistication. It looks like models that back-test beautifully because the last decade was kind. It looks like dashboards full of green lights. It looks like language that treats the absence of crisis as evidence of safety, instead of recognizing it as a warning sign that incentives are quietly drifting.

The markets reward this drift in the short run. When nothing goes wrong, caution looks like incompetence. The portfolio manager who holds extra cash underperforms. The bank that says "no" to marginal borrowers loses business to the bank that says "yes." The investor who refuses to buy the hot asset is told they "don't get it." In good times, safety is a social disadvantage. That is why forgetting is not merely personal; it is institutional. It is baked into competition.

Hyman Minsky captured the psychology with a phrase that sounds like an insult until you've watched it happen: *stability is destabilizing*. The idea is not mystical. When outcomes are stable, people take on structures that *require* stability to survive. They borrow short to invest long. They use leverage because it worked yesterday. They write contracts assuming liquidity will always be there. Over time, the system migrates from "we can pay our debts from our cash flows" to "we can refinance our debts because markets are friendly." That refinancing assumption is invisible during calm. It becomes obvious during a shock—usually at the worst possible moment.

The pandemic crash in March 2020 offered a modern proof of this logic. The global financial system had learned a lot since 2008. Banks, in many ways, were better capitalized. Yet stress still erupted—through nonbanks, through margin mechanics, through the sudden need for dollars, through the fragility of market liquidity. The cast changed. The plot did not. That's another reason forgetting is so persistent: the next crisis rarely repeats the last one perfectly. It rhymes. People who are trained to fight the previous war feel blindsided by the new one.

Institutional memory has an additional enemy now: technology's ability to rewrite the battlefield while we're still telling stories about the last fight. Viral attention can inflate a narrative-priced asset in days. Automated risk systems can force synchronized selling in minutes. New financial products can gather billions before anyone has a long enough dataset to understand their behavior under stress. Even if a veteran remembers what leverage did in 2008, that memory doesn't automatically tell them how leverage behaves when it's embedded in a different market structure, a different clearing system, or a different kind of collateral.

There's also the problem of *data gaps*, which is a polite term for "we don't know who owes what to whom until it matters." In tranquil periods, nobody wants to pay for boring plumbing reports. Institutions track what regulators require, what investors demand, and what executives can explain on an earnings call. The exposures that slip through are often the ones that are complex, cross-border, or technically outside the regulated perimeter. Then a shock arrives, and suddenly everyone is asking questions that should have been answered long before: Where is the leverage? Who is funding whom? What collateral is being reused? What happens if redemptions spike? The market becomes an emergency room triaging patients it has never met.

Regulators and policy makers suffer from their own version of forgetting. After a crisis, political will is high. New rules are written. Standards improve. Over time, lobbying intensifies, the crisis recedes, and the rules start to look "burdensome." The pendulum swings. It's not always cynical. Sometimes the system truly *is* safer and can afford to optimize. The danger is that optimization becomes a habit, and the habit becomes vulnerability.

Crypto regulation in the mid-2020s offered a live demonstration of uneven memory. Global bodies issued high-level recommendations emphasizing that "same risk" should be regulated similarly, regardless of the technology wrapper. That's institutional learning trying to keep up. But implementation across jurisdictions moved at different speeds, and gaps created incentives for activity to migrate to the most permissive venues. When rules are patchy, risk becomes nomadic. It doesn't disappear; it changes address.

In the United States, the GENIUS Act—signed on July 18, 2025—signaled another step in formalizing a stablecoin framework. The date matters because it shows how quickly memory can become policy after enough headlines pile up. Yet policy is not the same thing as safety. Laws still need rulemaking, supervision, enforcement, and coordination with the global system that stablecoins and exchanges inevitably touch. Institutional memory can write the rulebook; it cannot guarantee the players will obey it, or that the next game won't invent a new loophole.

The cycle of forgetting is also personal. The most dangerous moment for an individual investor is often not the first year of a bull market, when caution still feels reasonable. It's year six or seven, when the good times have matured into an identity. By then, your brain has created a private history in which "buying the dip" is not a tactic but a law of nature. You've seen it work enough times that you stop asking what would make it fail. You don't feel reckless. You feel experienced.

Markets are particularly good at training people the wrong way because they reward outcomes, not processes. If you take huge risk and get lucky, the result looks indistinguishable from skill. If you take prudent risk and underperform during a frenzy, the result looks indistinguishable from incompetence. Over a long enough career, most people will experience periods where luck flatters them and periods where prudence embarrasses them. The ones who survive learn to treat praise and shame as unreliable indicators of truth.

A subtle form of forgetting comes from the way institutions tell stories about success. When things go well, the narrative becomes, "We were disciplined," even if leverage quietly crept up. When things go badly, the narrative becomes, "It was unprecedented," even if the warning signs were familiar. These stories are not always lies. They are just incomplete. And incomplete stories are a hazard, because they teach the next generation the wrong lessons.

So what does survival look like in a world where bubbles are inevitable? It doesn't look like permanent pessimism. Pessimism has its own trap: if you sit out every rally because you're waiting for a perfect valuation and a guarantee of safety, you can miss the compounding that actually builds wealth. Survival is not about avoiding all risk. It's about refusing to take the kind of risk that ends your participation.

That leads to an unglamorous but liberating principle: you don't need to be right about the next bubble to be safe from it. You need a portfolio and a financial life that can tolerate being wrong in the short run. The crowd's favorite move is to demand all-or-nothing thinking. Either you're a believer or a hater. Either you're "all in" or you "don't get it." Survival lives in the middle: partial participation, deliberate sizing, and the humility to admit that the future doesn't owe you a smooth path.

Humility sounds soft until you realize it is a hard constraint on behavior. Humility is what stops you from using leverage because you've "never been wrong." Humility is what keeps you from treating liquidity as guaranteed. Humility is what makes you diversify even when concentration would feel heroic. In bubbles, humility is not a personality trait. It's a risk control.

Institutional memory can be built, but it has to be built on purpose, like fire codes. One practical method is to formalize "stories of failure" inside organizations: post-mortems that are specific, honest, and preserved. Not the sanitized versions that protect reputations, but the useful ones that protect future balance sheets. Another is to rotate people across roles so knowledge of market plumbing doesn't become the private property of a few specialists. A third is to stress-test for narratives, not just numbers: what happens if your biggest position becomes socially untouchable to sell? What happens if an app-driven crowd pushes volatility beyond your assumptions? What happens if a stablecoin you treat as cash stops being cash?

For individuals, the parallel is simpler: build your own memory before the market forces you to. Write down what you believe when you are calm, because you won't have the same beliefs when you are panicking or euphoric. The future version of you—the one staring at a violently moving chart—will not be impressed by your current intelligence. They will be grateful for your current discipline.

A surprisingly effective habit is to keep a "risk journal" with dated notes. When you make an investment, record why you bought, what would make you sell, and what you think could go wrong. Then, when the world changes, update the journal rather than updating the story in your head. This is how you prevent hindsight from quietly rewriting your identity as "someone who always saw it coming." Hindsight is emotionally soothing and financially lethal.

Another survival habit is to treat liquidity like oxygen. When it's plentiful, you don't notice it. When it's scarce, it's all you can think about. Keeping a genuine cash buffer, avoiding forced-selling scenarios, and refusing personal leverage that can trigger a margin-like spiral in your own life (credit card balances, fragile income assumptions, overcommitted fixed expenses) is not conservative. It's adaptive. It's how you stay in the game long enough for compounding to matter.

And if there's one final truth worth holding close, it's that bubbles are not only about greed. They are also about hope. Many people chase manias because their baseline financial reality feels hopeless: wages that don't keep up, housing that feels unreachable, a retirement math problem that seems unsolvable without extraordinary returns. That doesn't make bubbles good. It makes them understandable. A humane approach to financial survival acknowledges that emotion rather than denying it. You can respect the feeling without obeying it.

Actionable takeaway: create a personal "institutional memory" page and revisit it twice a year. On it, write:

- your maximum allowable loss in any single position,
- your rule against using leverage for speculation, and
- your minimum liquidity buffer (in months of expenses).

Sign and date it. When the next frenzy arrives—and it will—the goal is not to outsmart the crowd in real time. The goal is to make sure the crowd cannot talk you into breaking the rules that keep you solvent.

www.ingramcontent.com/pod-product-compliance
Lightning Source LLC
LaVergne TN
LVHW022213200926
848481LV00006B/1953